Biochemical Testing - Clinical Correlation and Diagnosis

Edited by Varaprasad Bobbarala, Gaffar Sarwar Zaman, Mohd Nasir Mohd Desa and Abdah Md Akim

Published in London, United Kingdom

IntechOpen

Supporting open minds since 2005

Biochemical Testing - Clinical Correlation and Diagnosis
http://dx.doi.org/10.5772/intechopen.73769
Edited by Varaprasad Bobbarala, Gaffar Sarwar Zaman, Mohd Nasir Mohd Desa and Abdah Md Akim

Part of IntechOpen Book Series: Biochemistry, Volume 12
Book Series Editor: Miroslav Blumenberg

Contributors
Satish Ramanathan, Husniza Hussain, Rusidah Selamat, Fuziah Md Zain, Muhammad Yazid Jalaludin, Lim Kuang Kuay, Dr. Khushaboo Pandey, Om Prakash Mishra, Nataša Gros, Samreen Riaz, Muhamamd Sohail, Zilton Vasconcelos, Daniela Prado Cunha, Andrea Zin, Luiza Neves, Renata Campos Azevedo, Kheir Eddine Kerboua, Yana Sandlers, Diana Kitala, Wojciech Grzegorz Smętek, Jacek Węgrzyk, Agnieszka Klama-Baryła, Wojciech Łabuś, Małgorzata Kraut, Michał Szapski, Mariusz Nowak

First published in London, United Kingdom, 2020 by IntechOpen
IntechOpen is the global imprint of INTECHOPEN LIMITED, registered in England and Wales, registration number: 11086078, 7th floor, 10 Lower Thames Street, London, EC3R 6AF, United Kingdom
Printed in Croatia

British Library Cataloguing-in-Publication Data
A catalogue record for this book is available from the British Library

Additional hard and PDF copies can be obtained from orders@intechopen.com

Biochemical Testing - Clinical Correlation and Diagnosis
Edited by Varaprasad Bobbarala, Gaffar Sarwar Zaman, Mohd Nasir Mohd Desa and Abdah Md Akim
p. cm.
Print ISBN 978-1-78985-085-7
Online ISBN 978-1-78985-086-4
eBook (PDF) ISBN 978-1-78985-371-1
ISSN 2632-0983

For EU product safety concerns:
IN TECH d.o.o., Prolaz Marije Krucifikse Kozulić 3, 51000 Rijeka, Croatia, info@intechopen.com or visit our website at intechopen.com.

IntechOpen Book Series

Biochemistry

Volume 12

Varaprasad Bobbarala obtained a doctorate from Andhra University with specialization in Biochemistry, Medicinal Chemistry, and Microbiology. He is editor-in-chief, associate editor, reviewer, and editorial board member for dozens of high-impact international periodicals. He has authored/co-authored research and review articles in numerous peer-reviewed national and international journals in various subjects related to biomedicine, pharmacy, and microbiology. Dr. Bobbarala previously served as Chief Scientist of Research and Development (R & D) at Krisani Innovations Pvt. Ltd., before his current role the Chief Scientist and Director of Adhya Biosciences Pvt. Ltd., India. Currently, he is working in the areas of clinical diagnostics, antimicrobial resistance, drug discovery, production of commercially important chemicals by biotechnology routes, isolation of bio-active metabolites, and bio-efficacy studies.

Dr. Gaffar S. Zaman has an MD in Biochemistry from Assam Medical College & Hospital, Dibrugarh University, India; a Fellowship in Diabetes (FID) from Royal Liverpool Academy, United Kingdom; a Fellowship in Applied Nutrition (FIAN) from Medvarsity, Apollo Hospitals, India; and Post Graduate Diploma in Clinical Research (PGDCR) from Symbiosis University, India. He has almost 12 years of experience as an Assistant Professor at King Khalid Government University and Rajiv Gandhi University of Health Sciences. He is adept at quality development and curriculum design and is trained in e-learning methods. He has published more than 40 research papers in both national and international journals. He has also edited two books published by IntechOpen: Quality Control in Laboratory and Ultimate Guide to Insulin.

Mohd Nasir Mohd Desa is currently Associate Professor in the Department of Biomedical Science, Faculty of Medicine and Health Sciences, Universiti Putra Malaysia (UPM). He was a former head for Laboratory of Halal Science Research at Halal Products Research Institute, UPM. He holds a BSc in Microbiology from the University of Arizona, USA, and a MMSc and PhD in Medical Microbiology from the University of Malaya, Malaysia. He has authored/co-authored research articles in numerous peer-reviewed journals locally and internationally in various subjects related to biomedical science and food authentication. He is active in postgraduate research supervision and teaches undergraduate courses on Medical Microbiology and Laboratory Management to health science students.

Abdah Md Akim is Associate Professor in the Department of Biomedical Science, Faculty of Medicine and Health Sciences, Universiti Putra Malaysia. She holds a BSc in Biochemistry from the University of Cardiff, UK, and a MSc and PhD in Medical Biochemistry from the Universiti Putra Malaysia. She has written research articles on biochemistry, cancer, antioxidants, and inflammation in numerous peer-reviewed journals. She is involved in postgraduate research supervision and teaches undergraduate courses on Medical Biochemistry and Laboratory Management to health science students.

Editors of Volume 12:
Varaprasad Bobbarala
Adhya Biosciences, India

Gaffar Sarwar Zaman
Department of Clinical Laboratory Sciences, College of Applied Medical Sciences, King Khalid University, Abha, Kingdom of Saudi Arabia

Abdah Md Akim and Mohd Nasir Mohd Desa
University Putra Malaysia, Selangor, Malaysia

Book Series Editor: Miroslav Blumenberg
NYU Langone Medical Center, New York, USA

Scope of the Series

Biochemistry, the study of chemical transformations occurring within living organisms, impacts all of life sciences, from molecular crystallography and genetics, to ecology, medicine and population biology. Biochemistry studies macromolecules - proteins, nucleic acids, carbohydrates and lipids –their building blocks, structures, functions and interactions. Much of biochemistry is devoted to enzymes, proteins that catalyze chemical reactions, enzyme structures, mechanisms of action and their roles within cells. Biochemistry also studies small signaling molecules, coenzymes, inhibitors, vitamins and hormones, which play roles in the life process. Biochemical experimentation, besides coopting the methods of classical chemistry, e.g., chromatography, adopted new techniques, e.g., X-ray diffraction, electron microscopy, NMR, radioisotopes, and developed sophisticated microbial genetic tools, e.g., auxotroph mutants and their revertants, fermentation etc. More recently, biochemistry embraced the 'big data' omics systems.

Initial biochemical studies have been exclusively analytic: dissecting, purifying and examining individual components of a biological system; in exemplary words of Efraim Racker, (1913 –1991) "Don't waste clean thinking on dirty enzymes." Today however, biochemistry is becoming more agglomerative and comprehensive, setting out to integrate and describe fully a particular biological system. The 'big data' metabolomics can define the complement of small molecules, e.g., in a soil or biofilm sample; proteomics can distinguish all the proteins comprising e.g., serum; metagenomics can identify all the genes in a complex environment e.g., bovine rumen. This Biochemistry Series will address both the current research on biomolecules, and the emerging trends with great promise.

Contents

Preface

This book provides an overview of the analysis of blood, serum, plasma, urine, and tissues for a wide variety of substances such as substrates, enzymes, and hormones, and their use in diagnosis and monitoring of disease. The content provides qualitative and quantitative analysis of these biological fluids with their specific chemical constituents or physiologic processes.

The first chapter by Drs. Satish Ramanathan and Chakravarthy Narasimhachar Srinivas emphasizes the use of electrophoresis as an aid in diagnosing clinical conditions such as acute and chronic inflammations and various other diseases. Serum protein electrophoresis, which separates serum proteins into various fractions based on their molecular weight and electric charges, can provide specific measurements of various serum proteins such as alpha-1 antitrypsin (AT), alpha-1-chymotrypsin, thyroid-binding globulin, alpha-2 macroglobulin, haptoglobin, ceruloplasmin, transferrin, low-density lipoprotein, and immunoglobulins (IgG, IgA, IgM, IgD, and IgE) for diagnostic purposes.

The chapter by Husniza Hussain, Rusidah Selamat, Lim Kuang Kuay, Fuziah Md Zain, and Muhammad Yazid Jalaludin highlights the importance of urinary iodine measurement in diagnosing disease related to this marker.

The chapter by Khushaboo Pandey and Om Prakash Mishra Apart discusses sample analysis, collection, storage, and preparation. It also discusses various spectrophotometry, sensors, and chromatographic techniques.

The chapter by Nataša Gros covers the quality of blood collection tubes and examines the differences in tube characteristics of different brands and different lots. Because the tubes' attributes change over their shelf life, the author suggests fast, easy-to-perform testing procedures using purified water and low-cost equipment.

The chapter by Samreen Riaz and Muhamamd Sohail examines the use of extra virgin olive oil in reducing lipid profile in diabetes. It reports that this activity may be due to the oil's antioxidants. The authors use in silico bioinformatic analysis to screen and check the therapeutic potential of this plant.

The chapter by Zilton Vasconcelos, Renata Campos Azevedo, Andrea Zin, Luiza Neves, and Daniela Prado Cunha explores the advancement of serological diagnostic tools for Zika virus. Reverse transcriptase real-time PCR is suggested as the most reliable tool for Zika virus infection diagnosis. It detects viral RNA in both biological fluids and tissues and contributes to clinical case classification for initial description of developmental changes observed in neonates exposed congenitally to Zika virus.

A detail protocol of standardized hemolytic assay AP100 was elaborated in a chapter on Semi-Solid Phase Assay for Alternative Pathway Activity Assessment. The authors aimed to assist laboratories in the developing and low incomes countries by using this assay. The procedure is essentially the same as AP50 except the two items mentioned in the book chapter. The chapter by Kheir Eddine Kerboua emphasizes the many advantages of AP100 over AH50, where AP100 has at least nine applications in disease diagnosis and follow-up.

The chapter by Yana Sandlers gives an overview of amino acid metabolism-related inherited disorders. It presents amino acid analysis for the diagnosis and routine monitoring of this category of inborn errors of metabolism.

Finally, the chapter by Wojciech Smętek, Jacek Węgrzyk, Agnieszka Klama-Baryła, Wojciech Łabuś, Małgorzata Kraut, Michał Szapski, Mariusz Nowak, and Diana Kitala discusses the importance of performing good quality control in accordance to good manufacturing practice to produce high-quality, advanced therapy medicinal products.

The Editors would like to acknowledge all chapter contributors for their excellent pieces of writing and IntechOpen for publishing this book.

Varaprasad Bobbarala
Adhya Biosciences,
India

Gaffar Sarwar Zaman
Department of Clinical Laboratory Sciences,
College of Applied Medical Sciences,
King Khalid University,
Abha, Kingdom of Saudi Arabia

Abdah Md Akim and Mohd Nasir Mohd Desa
University Putra Malaysia,
Selangor, Malaysia

Chapter 1

Serum Protein Electrophoresis and Its Clinical Applications

Satish Ramanathan and Chakravarthy Narasimhachar Srinivas

Abstract

This chapter focuses on the principle of electrophoresis and its utilization in a clinical laboratory. A sincere attempt has been made to discuss about clinical applications of serum protein electrophoresis, throwing light on the significance of serum protein electrophoresis in the management of multiple myeloma. Emphasis has been made on quality assurance in terms of accuracy and precision in electrophoresis to ensure reliability of patient results. A note on issues with lack of standardization of reporting of electrophoresis and an insight into global efforts to standardize the reporting of the assay has been included in this chapter.

Keywords: electrophoresis, gamma globulins, polyclonal, oligoclonal, biclonal, myeloma

1. Introduction

Serum protein electrophoresis is an electrophoretic method of separating proteins present in the serum to various fractions based on their molecular weight and electric charges. Electrophoresis had been widely used in clinical medicine for aiding in diagnosis of various clinical conditions like acute and chronic inflammations, monoclonal gammopathies, nephropathy, liver diseases, etc. This chapter discusses the clinical applications of serum protein electrophoresis [1] including the quality control practices and its implications [2].

2. Principle

The separation of proteins by electrophoresis is based on the fact that charged molecules usually migrate through a matrix/medium upon application of an electrical field [3]. The rate at which proteins move in an electric field is determined by a number of factors of the electrophoretic system and the nature of proteins itself. Some factors to mention are the strength of the electric field, temperature of the system, pH of the ions, concentration of buffer etc. [4]. Proteins vary in their size and shape and have the charges determined by the dissociation contents of their amino acids. Smaller proteins usually migrate faster, and larger proteins take a longer time. This physical property of proteins is exploited for its separation by employing the electrophoretic technique.

The most commonly employed variant of electrophoresis for serum protein separation is zone electrophoresis in which the serum proteins are separated into zones or

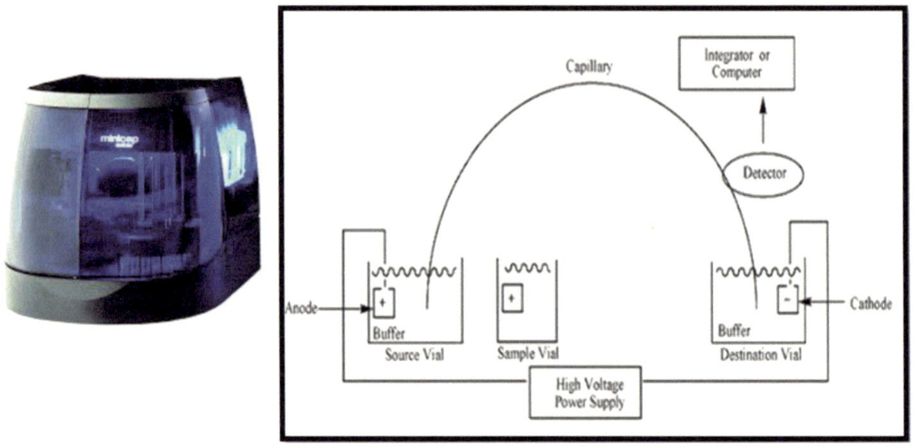

Figure 1.
Sebia Minicap flex piercing capillary electrophoresis.

fractions and interpreted accordingly [5]. There are several support mediums available for separation of serum proteins including agarose, cellulose acetate, capillary medium etc. [6]; when a capillary medium is used, the technique is known as capillary zone electrophoresis (CZE). Capillary electrophoresis is the preferred method when compared to its competitors including agarose gel electrophoresis due to the following reasons. CZE provides an improved resolution due to the following factors:

a. The use of "electroendosmosis" principle which improves the resolution of separation

b. Employing a "high-voltage" electric current which aids in improving the throughput (the processing time) and the resolution of protein separation.

Below is an illustration of capillary electrophoresis (Sebia Minicap Flex Piercing) (**Figure 1**). Sebia Minicap Flex Piercing capillary electrophoresis works on the principle of capillary electroendosmosis under high-voltage electric current. The Flex Piercing model of Sebia CZE aids in testing of human blood with capped tubes which in turn eliminates the biohazard xassociated with handling of uncapped samples.

3. Revisiting the basics: an insight into the protein family

Serum proteins are a family of albumin and globulins. Albumin is the major fraction synthesized from human liver endogenously and available through various dietary sources exogenously including egg, meat, pulses, milk etc. Globulins are a group of proteins subclassified into alpha-1, alpha-2, beta-1, beta-2, and gamma globulins based on the electrophoretic mobility (**Figure 2**). The normal biological interval of serum total proteins in a healthy adult ranges between 6 and 8 g/dl which includes Serum Albumin: 3.5–4.5 g/dl and Globulins: 2.5–3.5 g/dl.

3.1 Albumin

Albumin is a 69 kDa protein. It is the most abundant protein in serum. Albumin is synthesized in the liver and functions as a transport protein of various substances like bilirubin, enzymes, hormones, drugs etc. It also maintains fluid volume within the vascular space. Albumin is the first protein fraction to appear near the anode in

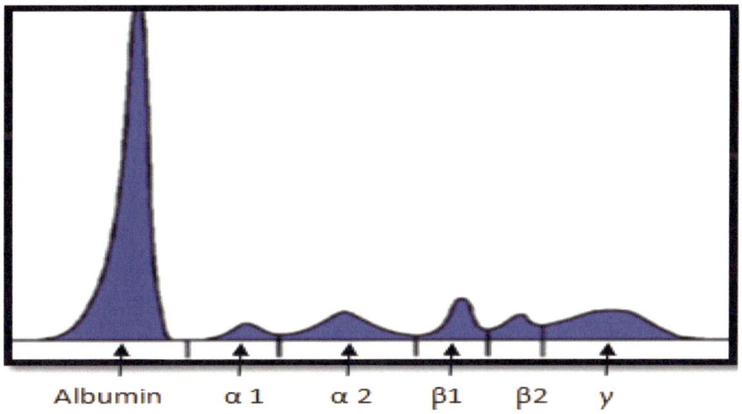

Figure 2.
Serum protein family with fractions.

SPE. Altered levels of serum albumin are associated with various clinical conditions. Low levels of albumin are clinically significant and are termed as hypoalbuminemia.

Decreased concentration of serum albumin (hypoalbuminemia) indicates either a poor dietary intake (malnutrition) or a decreased production or an increased loss. Chronic liver disease is a common clinical condition associated with decreased albumin production, and chronic kidney disease (CKD) is the most common disease associated with an increased loss of albumin in urine (proteinuria). This clinical condition is otherwise known as nephropathy. Other causes of hypoalbuminemia include acute and chronic inflammation, critical illness, pregnancy etc.

Apart from hypoalbuminemia which is commonly observed in an electrophoretogram, there are a few variations which can be observed in the in albumin peak including:

a. Bilirubin, Triglycerides if present in high levels in serum may appear as a blunt peak which is seen adjacent to the cathode near the albumin peak.

b. Prealbumin (transthyretin)—increased levels of pre albumin, if present due to various clinical conditions including several inflammatory diseases is seen as a blunt anodal peak distinctly separated from the peak of albumin.

c. A rare variant observed in the albumin peak is bisalbuminemia which is a rare condition, with no clinical features, in which the serum contains two albumin variants of different electrophoretic mobilities, usually in equal concentrations, though the total concentration of albumin is normal. Bisalbuminemia may be hereditary or acquired. The acquired type has been more frequently reported in chronic renal disease and pancreatitis and in patient with chronic renal disease. Two (rather than one) albumin bands may represent bisalbuminemia. Hereditary condition is a rare anomaly caused by a genetic lesion in the albumin gene usually a point mutation.

d. Analbuminemia (absence of albumin) is another genetically inherited metabolic disorder and was first described in 1954. This disorder is rare and affects less than 1 in 1 million births.

The most important aspect of such albumin variants lies in quantification of an albumin peak in such scenarios followed by interpretation and clinical correlation (**Figure 3**).

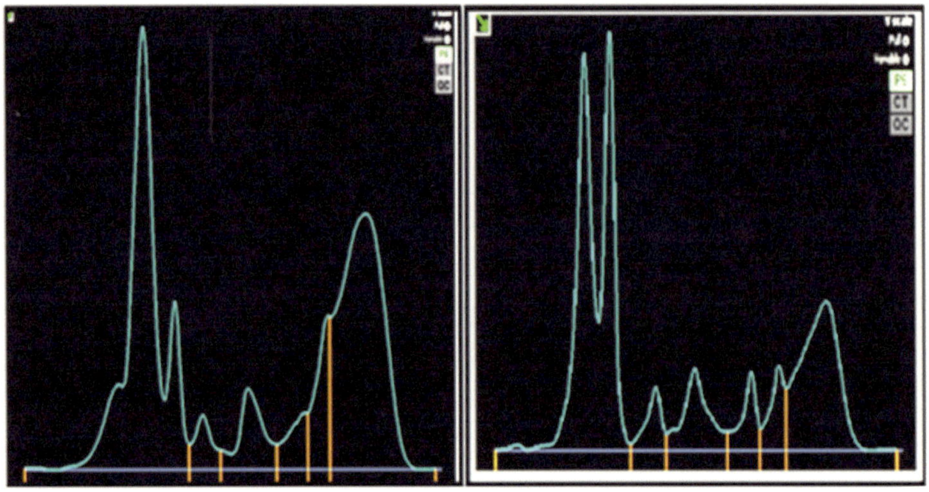

Bilirubin interference (blunt cathodal peak adjacent to albumin

Bisalbuminemia (bifid peaks in albumin region)

Figure 3.
Abnormal electrophoretic patterns of albumin zone.

3.2 Alpha fraction

As electrophoresis proceeds toward the negative portion of the gel (cathode), the alpha zone is the next band after albumin. The alpha zone is subdivided into two zones: the alpha-1 peak and alpha-2 peak.

The alpha-1 peak consists of alpha-1 antitrypsin (AT), alpha-1-chymotrypsin, and thyroid-binding globulin. Alpha-1 antitrypsin is an acute-phase reactant. The concentration of alpha-1 antitrypsin increases in conditions of inflammation and is usually decreased in patients with alpha-1 antitrypsin deficiency or decreased production of globulin in patients with severe liver disease. A rare variant of alpha-1 antitrypsin is encountered occasionally characterized by a split peak pattern of alpha-1 globulins.

The alpha-2 peak consists of alpha-2 macroglobulin, haptoglobin, and ceruloplasmin. Alpha-2 macroglobulin accounts for about 3% of the total protein in the serum. Because of the variable migration of the haptoglobin types, a2-macroglobulin is often adjacent to, or co migrating with, haptoglobin and is therefore not seen as a discrete band.

A distorted pattern of alpha-2 region in electrophoresis is seen commonly in conditions of hemolysis, including in vivo and in vitro. The pathophysiology behind this pattern is the formation of hemoglobin-haptoglobin complexes in these conditions. This is a physiological adaptive response by human physiology to conserve hemoglobin released as a result of RBC breakdown into circulation and hemoglobin being a smaller globular protein is bound to be lost in urine. Hence to preserve it, haptoglobin is consumed to form complex with hemoglobin which results in the formation of a macromolecular protein which is retained in circulation making hemoglobin available for the production of RBCs and prevention of anemia.

Haptoglobin and ceruloplasmin are acute-phase reactants, and hence increased in acute inflammatory states.

Alpha-2-macroglobulin is increased in nephrotic syndrome and cirrhosis of the liver. The elevation of alpha-2 macroglobulin is distinctly evident in nephritic syndrome, since it is a bulky molecule, and hence retained in circulation to compensate for the loss of other proteins in urine which is evident in form of proteinuria in urine microscopic examination.

Ceruloplasmin is an important copper-binding transport protein produced by the liver. Ceruloplasmin concentrations are markedly decreased in conditions of Wilson's disease. The disadvantage of serum protein electrophoresis is that it will not aid in the detection of a decreased ceruloplasmin.

3.3 Beta fraction

The beta zone usually is subdivided into two peaks, beta-1 and beta-2 in CZE. Beta-1 zone comprises proteins like transferrin and low-density lipoprotein (LDL).

Transferrin functions to transport non-heme ferric iron from the gastrointestinal tract. Each Transferrin molecule can bind two molecules of free iron. An increased beta-1 band is observed in iron deficiency anemia due to an increased level of free transferrin and also in pregnancy. Determinations of the transferrin levels are useful in distinguishing between iron deficiency anemia (inadequate intake or chronic hemorrhage with loss of iron stores) and hemolytic anemia, in which transferrin levels are low resulting in a beta-1 peak of low amplitude. Transferrin is usually decreased in alcoholic cirrhosis. Transferrin is also decreased during renal disease and thermal injuries.

The beta-2 band is mostly composed of complement proteins, C3 and C4. Elevated beta-2 zone can be caused in inflammatory states due to activation of complement cascade which include C3 and C4 too.

A reduced beta-2 peak intensity can be encountered in an aged sample, since the immune complexes are used up and low serum levels of complements are evidenced.

Fibrinogen is a protein with molecular weight of 340 kDa protein. Sometimes a small fibrinogen band can be seen in serum protein electrophoresis due to the insufficient clotting or failure to remove the serum from the clot. This fibrinogen band is seen between beta-1 and beta-2 regions. This band is also seen in patients who are receiving heparin therapy. It is also an important indicator of the sample type being analyzed. When plasma is used in the place of serum for protein electrophoresis, fibrinogen present in plasma appears in the beta-2 region, and this has the potentiality to interfere with the detection of monoclonal gammopathies in such patients (**Figure 4**).

3.4 Gamma fraction

One of the main clinical implications of serum protein electrophoresis is to aid diagnosis of disorders associated with alterations of gamma globulins. Gamma region comprises mainly of serum immunoglobulins. The five major classes of immunoglobulins are IgG, IgA, IgM, IgD, and IgE. The immunoglobulins are characterized by the presence of two protein moieties named as heavy chain and light chain. The classification of immunoglobulin had been made based on the composition of heavy chains, while the light chains are of two types including kappa or lambda. Physiologically, kappa forms the major light chain fraction among the two.

Various clinical conditions are associated with alteration of gamma globulins including:

a. Hypergammaglobinemia (increased serum gamma globulin levels)

b. Hypogammaglobinemia (decreased serum gamma globulin levels)

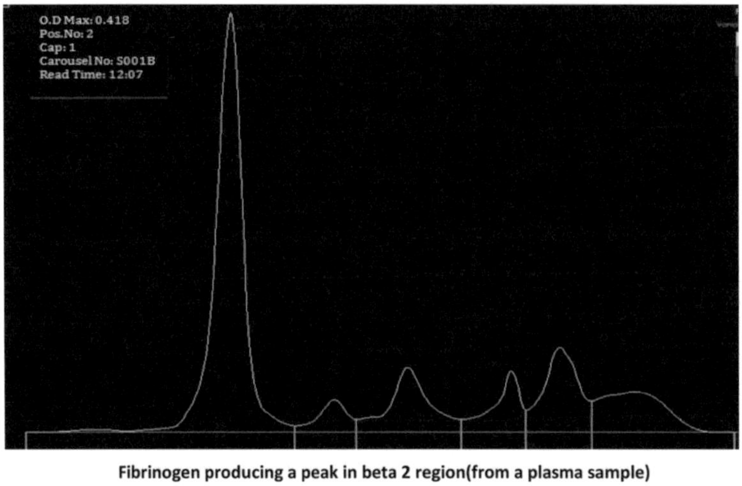

Figure 4.
Fibrinogen producing a peak in beta 2 region (from a plasma sample).

Hypergammaglobinemia (gammopathies):

Gammopathy is defined as abnormal proliferation of the lymphoid cells producing immunoglobulins. There are four types of gammopathies: polyclonal, monoclonal, biclonal, and oligoclonal.

Polyclonal gammopathies are defined as heterogeneous increase in immunoglobulins involving more than one cell line, commonly caused by a variety of inflammatory conditions (chronic inflammation), infections, chronic liver diseases (cirrhosis), chronic kidney diseases, etc.

Monoclonal gammopathies are characterized by a homogenous increase produced by clonal population of mature B cells, most commonly plasma cells. Monoclonal immunoglobulins seen in these conditions are also known as Para proteins. The classic interpretative terminology used in clinical laboratory medicine for describing a monoclonal immunoglobulin in SPE is “M” band where M stands for monoclonal. Common clinical disorders producing “M” Band in SPE include multiple myeloma and plasmacytoma in usually 60% of cases and Waldenströms Macroglobulinemia, lymphomas, and leukemia in approximately 10% of cases. Certain monoclonal gammopathies produce “M” band in electrophoretic regions other than in gamma regions, commonly being beta region especially in case of IgA and IgG myeloma.

Biclonal gammopathies are characterized by a double peak in the gamma region. This electrophoretic pattern is seen when there is a biclonal proliferation of immunoglobulins encountered in multiple myeloma. A biclonal pattern is also seen in monoclonal gammopathies associated with IgA and IgG. In such scenarios, these immunoglobulins appear as polymerized and monomerized forms which elute as biclonal peaks in gamma region or in beta region, respectively (**Figure 5**).

The oligoclonal pattern of gamma region is characterized by more than two peaks evident in the gamma region. This pattern is commonly seen in autoimmune disorders, light chain myelomas (characterized by clonal proliferation of light chains), amyloidosis, etc. (**Figure 5**).

Apart from serum immunoglobulin, C-reactive protein (CRP) also is evident in the gamma region. C-reactive proteins levels usually increase during inflammatory responses.

Apart from the common causes of altered electrophoresis picture specific to the particular zones, a sharp distinct peak when evident especially in beta or alpha region

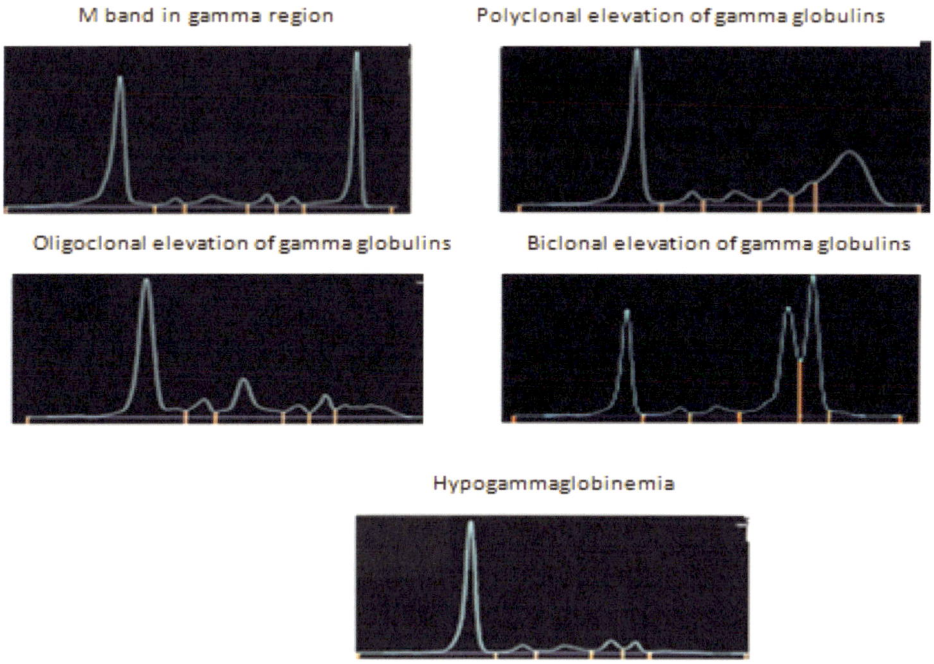

Figure 5.
Abnormal electrophoretic patterns of gamma zone.

should raise a high index of diagnostic suspicion of multiple myeloma since a few monoclonal immunoglobulins shall migrate in these zones too, in contrary to the classical gamma zone M protein pattern, which is commonly reported in these conditions.

4. Role of SPE in multiple myeloma work-up

According to the International Myeloma Foundation, plasma cell dyscrasias are group of plasma cell disorders involving a wide spectrum of pathologies including:

1. MGUS—monoclonal gammopathy of undetermined significance
2. MGRS—monoclonal gammopathy of renal significance
3. Smoldering myeloma
4. Multiple myeloma (which includes various subtypes including nonsecretory myeloma (NSMM), light chain myeloma, secretory multiple myeloma)

Criteria for diagnosis and differentiation of the plasma cell disorders based on International Myeloma Working Group (IMWG) criteria:

1. **Monoclonal gammopathy of undetermined significance** (MGUS)
    - M protein (Monoclonal band)—<3 g/dl
    - Bone marrow biopsy—<10% plasma cells seen

- No clinical symptoms/signs
- Normbal free light chain ratio in serum

2. **Monoclonal gammopathy of renal significance (MGRS)**
    - M protein (monoclonal band)—<3 g/dl
    - Bone marrow biopsy—<10% plasma cells seen
    - Renal disease with elevated serum creatinine
    - Normal Free light chain ratio in serum
3. **Smoldering myeloma**
    - M protein (monoclonal band)—<3 g/dl
    - Bone marrow biopsy—>10% plasma cells seen
    - Abnormal free light chain ratio in serum
    - Clinically significant. Clinical diagnosis includes a tetrad of "ÇRAB" which stands for (one of the four shall be present):
        - C—hypercalcemia
        - R—renal abnormalities (elevated creatinine)
        - A—anemia
        - B—bone lesions
4. **Multiple myeloma**
    - M protein (monoclonal band)—>3 g/dl
    - Bone marrow biopsy—>10% plasma cells seen
    - Abnormal free light chain ratio in serum
    - Clinical diagnosis includes a tetrad of "ÇRAB" which stands for (one of the four shall be present):
        - C—hypercalcemia
        - R—renal abnormalities (elevated creatinine)
        - A —anemia
        - B—bone lesions

There are exceptions in SPE findings in certain cases of multiple myeloma wherein the SPE does not reveal any significant alteration or a clue toward the diagnosis.

These variants of multiple myeloma characterized by an abnormal bone marrow (increased plasma cells) but a normal SPE are termed as nonsecretory myelomas which account to 1–2% of multiple myelomas. In such cases, an immunoassay of free light chains (FLC) in serum provides a diagnostic clue toward NSMM which show a significant disproportionate elevation of usually a clone of light chains (kappa or lambda) with an alteration in kappa/lambda ratio (normal Ratio is between 0.60 and 1.65). A commonly encountered phenomenon with laboratory testing of FLC includes "prozone" effect or "hook" effect which occurs due to antigen excess and requires appropriate dilution to obtain reliable results.

Bence-Jones protein estimation in urine is an antique piece of laboratory evidence toward multiple myeloma, which is characterized by detection of light chains in urine. But since the methodology of testing is manual and does not provide standardization, this has been replaced by urine FLC analysis in laboratories practicing good clinical laboratory practices (GCLP).

One more valiant laboratory investigation which is an essential requisite for multiple myeloma work-up includes immunoelectrophoresis.

One common principle employed in immunoelectrophoretic technique involves the use of specific antihuman immunoglobulins (e.g., Anti-IgG, Anti-IgA, Anti-Kappa, etc.) as a preprocessing step which results in precipitation of immunoglobulins if present and disappearance of the band/peak contributed by that specific immunoglobulin. Hence this technique is also known as immunosubtraction. This technique aids in typing the specific type of immunoglobulin (including the type of light chain) contributing to myeloma. This technique is supplemented by quantification of serum immunoglobulins by an immunoassay.

4.1 SPE and its clinical significance

SPE is a semiquantitative investigation which involves technical expertise to recognize the specific electrophoretic patterns and associate with various clinical conditions. This requires a laboratory practice integrated across various divisions of laboratory and with respective clinical and ancillary divisions of clinical medicine [1].

With respect to SPE, the laboratory professionals shall act as consultants to the clinical consultants. This is possible in scenarios where the clinician does not arrive at a provisional diagnosis of a gammopathy and the laboratory picks up the diagnostic clue toward gammopathy through an increased serum total protein level (<8 g/dl) and an altered serum albumin globulin (AG) ratio (which is usually altered in gammopathy). A normal AG ratio ranges between 1.2 and 1.8, while there is a significant reduction in the ratio in patients with gammopathy. This becomes an incidental finding which leads to a concept of "reflex" testing for multiple myeloma work-up including SPE, upon consent from the treating clinician and the patient.

These are some of the common SPE patterns associated with various clinical conditions:

- **Inflammation:** Increased intensity of alpha-1 and alpha-2 with a sharp leading edge of alpha-1 may be observed, but with chronic inflammation the albumin band may be decreased with increased gamma zone due to the polyclonal gammopathy.

- **Nephrotic syndrome:** The albumin band is decreased due to hypoalbuminemia. In addition, the alpha-2 band may be more distinct.

- **Cirrhosis or chronic liver disease:** A low albumin band due to significant hypoalbuminemia with a prominent beta-2 band and beta-gamma bridging is a characteristic feature. In addition, polyclonal hypogammaglobinemia is observed.

- **Malnutrition:** Decreased albumin levels [1].
- **Alpha-1 antitrypsin deficiency**: Inflammatory condition, pregnancy.
- **Hemolysis:** Altered electrophoretic pattern of small indistinct peaks in alpha-2 region.

4.2 Quality assurance in SPE

Quality assurance in SPE is an essential prerequisite to ensure reliability of an SPE result [2]. There are two major aspects of analytical quality including precision (measure of precision) and accuracy (measure of trueness).

Good clinical laboratory practices demand processing of an internal quality control (IQC) for assessment of precision and external quality assurance (EQA)/ proficiency testing (PT testing) for accuracy assessment. IQC is a material which can be prepared in house (patient sample) or available commercially and is to be processed before a patient sample is taken up for processing.

The clinical laboratory has its responsibility to select and use an IQC which has a matrix comparable to patient sample, preferably covering the clinical decision point (cut off value that differentiates between a normal and abnormal result). EQA is an external assessment of the analytical quality wherein the laboratory processes a blinded sample and the results are compared against a reference method and/or against the consensus value of other participant laboratories for that specific sample.

The laboratory has to hold responsibility in selecting a suitable EQA provider who shall preferably be accredited to ISO 17043. If an EQA program is not available, the laboratories shall participate in exchange of samples with referral laboratories with a similar methodology and a comparable quality of testing standard.

5. Reporting of results and its standardization

Reporting SPE requires interpretation of the electrophoretic pattern which is followed by comments of such an interpretation along with the piece of advice to the clinician if indicated. There is a big lacuna in the format of reporting of SPE, each laboratorian using his/her own means of interpreting and communicating. It is the need of the hour to have a standardized format of reporting SPE for ensuring patient safety and clinician follow-up. There are no international guidelines, though the working party on standardized reporting of protein electrophoresis which is an initiative of the Australasian Association of Clinical Biochemists has come out with a standardized format of reporting SPE.

6. Conclusion

In the current scenario, it becomes the responsibility of each and every laboratory to ensure that all relevant information is available in a SPE report, easily read, understood, and interpreted by a clinician. This becomes the core of a clinical laboratory practice.

Acknowledgements

We would wish to acknowledge and thank the management of MIOT hospitals for providing us with the infrastructure and technology to explore, learn, and contribute to our patients' well-being.

We would wish to thank our technical staff, Mr. Mathivanan Durairaj, and his team for their invaluable contribution by sharing clinical cases.

We wish to acknowledge Trivitron technologies for their valuable support in installation and continual service and application support for Sebia CZE.

Author details

Satish Ramanathan[1*] and Chakravarthy Narasimhachar Srinivas[2]

1 Division of Clinical Biochemistry, MIOT Hospitals, Chennai, Tamil Nadu, India

2 Laboratory Medicine, MIOT Hospitals, Chennai, Tamil Nadu, India

*Address all correspondence to: satishraksha1980@gmail.com

References

[1] O'Connell TX, Horita TJ, Kasravi B. Understanding and interpreting serum protein electrophoresis. American Family Physician. 2005;**71**(1):105-112

[2] Jenkins MA. Electrophoresis. Quality Control and Quality Assurance Aspects of the Routine use of Capillary Electrophoresis for Serum and Urine Protein in Clinical Laboratories. Wiley Online Library. Jun 2004;**25**(10-11):1555-1560

[3] Jenkins MA. Serum Protein Electrophoresis. Clinical Applications of Capillary Electrophoresis. Molecular Biotechnology; 2000. pp. 11-19

[4] Ninfa AJ, Ballou DP, Benore M. Fundamental Laboratory Approaches for Biochemistry and Biotechnology. Hoboken, NJ: Wiley; 2009. p. 161

[5] Smith I, editor. Zone Electrophoresis: Chromatographic and Electrophoretic Techniques. 4th ed. Elsevier Ltd.; 1976

[6] Tate J, Caldwell G, Daly J, Gillis D, Jenkins M, et al. Recommendations for standardized reporting of protein electrophoresis in Australia and New Zealand. Annals of Clinical Biochemistry. 2012;**49**:242-256

Chapter 2

Urinary Iodine: Biomarker for Population Iodine Nutrition

Husniza Hussain, Rusidah Selamat, Lim Kuang Kuay, Fuziah Md Zain and Muhammad Yazid Jalaludin

Abstract

Many reports or manuals had focused on the implementation of iodine deficiency disorder (IDD) elimination programme from the point of view of the programme managers. In this chapter, we will focus on the importance of urinary iodine testing, its related diagnosis and further biomarker testing suggested for further diagnosis related to thyroid health. This chapter will be relevant for the respondents to the monitoring programme, particularly the 8–10-year-old schoolchildren and pregnant women, i.e., the vulnerable targeted groups from either the iodine-deficient areas or the Universal Salt Iodization (USI) gazetted areas. USI has been proposed by the World Health Organization (WHO) as the most cost-effective programme to eliminate IDD, and it is also a way to increase the intelligent quotient (IQ) of the world population for the future. This chapter had been laid out so that the readers will know briefly the rationale behind the testing of urinary iodine among schoolchildren and pregnant women under the implementation of the USI programmes in their countries and their benefits, especially the utilisation of urinary iodine as the biomarker to portray the population iodine status. Diagnosis including iodine-induced thyroid diseases and further biomarkers measurement besides urinary iodine is also discussed briefly.

Keywords: urinary iodine, biomarker, iodine nutrition, population, thyroid status

1. Introduction

1.1 Importance of urinary iodine testing to determine population iodine nutrition

All iodine in the blood is in the iodide form either it is taken up by the thyroid and converted into thyroid hormone or being excreted in the urine. Almost 90% of the ingested iodine is excreted in the urine. Therefore, urinary iodine excretion is a good biomarker of very recent dietary iodine intake [1]. On an individual basis, 24-hour urine sample is necessary for the assessment of iodine intake as the level is more consistent in iodine-deficient populations than in those with adequate iodine intake (**Figure 1**). On a population basis, the median urinary iodine concentration (mUIC) of spot urine from sufficiently large randomly selected 8–10-year-old children or adults has been shown to provide useful information on the average iodine intake or status of a community. On an individual basis, urinary iodine varies from day to day and even within a given day. However, this variation tends to even out among population [2]. Most of the epidemiological IDD studies had emphasised on rapid inexpensive methods of urinary iodine determination that could be applied to a large number of samples [3].

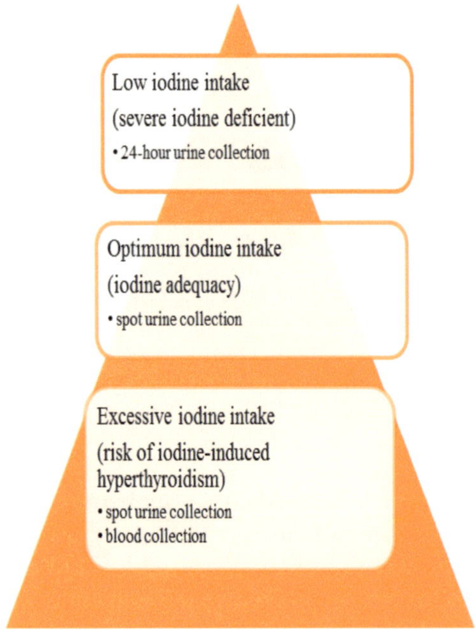

Figure 1.
Degrees of iodine intake (iodine nutrition status) and their suitable types of human sample collection to determine the median urinary iodine concentration (mUIC).

The main biochemical indicator that is widely used for the assessment of IDD is urinary iodine concentration (mUIC) [4]. The advantages of mUIC as an indicator of IDD are that the method directly reflects iodine supply of the individual, it is objective and non-invasive and urine samples can be kept for later analysis. However, the disadvantages of this method are that it requires laboratory space, special facilities and skilled technician to provide accurate determinations. In addition, this method reflects only current but not past intake of iodine [5].

Epidemiological studies stated that the population distribution of urinary iodine is required rather than individual levels. The frequency distribution of urinary iodine usually skewed towards elevated values; hence, the median value is considered instead of the mean as indicating the status of iodine nutrition [1]. The mUIC of 100 μg/L and above defines a population which has no IDD; i.e. at least 50% of the sample should be >100 ug/L. In addition, not more than 20% of sample should be below 50 μg/L. Iodine nutrition status is based on six categories of urinary excretion classification (**Table 1**) [3].

Median urinary iodine concentration (μg/L)	Severity of IDD
<20	Severe deficient
20–49	Moderate deficient
50–99	Mild deficient
100–199	Optimal
200–299	More than adequate
>300	Excessive
Source: Ref. [3].	

Table 1.
Epidemiological criteria for assessing IDD in a population based on median urinary iodine concentration.

1.2 Iodine role in thyroid hormone synthesis and its contribution to human body

1.2.1 Thyroid hormone synthesis

Iodine is grouped under micronutrients, and it is needed in small amount, but it is very important for the development of optimum human growth. Iodine is needed in the synthesis of thyroid hormones [6]. Through iodination, one, two, three or four iodine atoms are bound to tyrosine to form monoiodothyronine (MIT), diiodothyronine (DIT), triiodothyronine (T_3) or thyroxine (T_4), respectively, through the action of iodinase enzyme. Iodine is absorbed from the gastrointestinal system, will enter the blood circulation and will be transported into the thyroid follicle cells through the sodium/iodine (Na/I) symporter. In iodide form, it will then be transported to the thyroid follicle colloid through pendrin. Concurrently, thyroglobulin (TG) is being synthesised in the endoplasmic reticulum (ER) and being secreted into the follicle colloid through exocytosis. TG is the transporter protein of the thyroid hormones in the thyroid follicle colloid. It consists of branches of tyrosine molecules which will then be bound to iodine through the iodination process, forming the MIT and the DIT. When one MIT and one DIT bind, T_3 will be formed, while upon binding of two DITs, T_4 will then be formed. These TG-bound thyroid hormones will enter the thyroid follicular cell through endocytosis. TG will then undergo proteolysis, and T_3 and T_4 will be transported into the blood circulation through the MCT symporter [7] (**Figure 2**).

1.2.2 Iodine's contribution to human body

Iodine is a micronutrient which is present in the body in minute amount. The quantity of iodine required by an individual is about 150–200 μg/day [8]. Its main role is in the synthesis of thyroid hormone which is essential for the brain

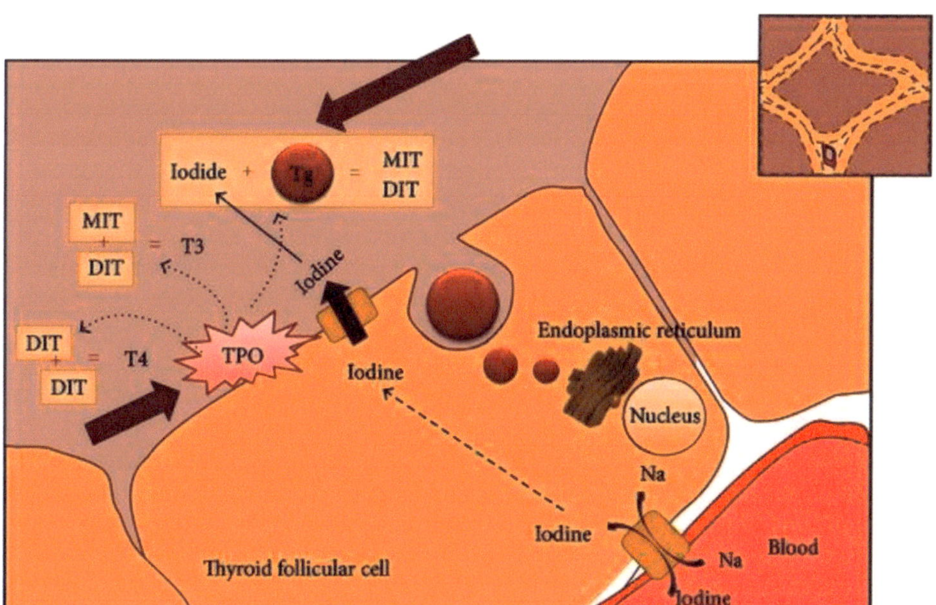

Figure 2.
Thyroid hormone synthesis in the thyroid gland (image reproduced with permission from the rights holder, Journal of Thyroid Research) [13].

and physical development [9]. The regulation of thyroid hormones is under the control of the pituitary gland through thyroid-stimulating hormone (TSH). TSH secretion is regulated by a 'feedback' mechanism related to the level of thyroid hormones thyroxine [3,5,3′,5′-tetraiodothyronine (T_4)] in the blood. Iodine is needed in the human body, and as the blood T_4 falls, the pituitary TSH secretion is increased. In severe iodine deficiency disorders (IDD), the level of T_4 remains lowered, and the level of TSH remains elevated. Both these measurements are used for diagnosis of hypothyroidism due to IDD at various stages in life particularly in neonates [10].

2. Iodine deficiency disorders (IDD) and symptoms

2.1 Causes of IDD

Most of the iodine exists in the ocean and seafood, including saltwater fish, shellfish, kelp, seaweed and seaweed products which can provide a considerable amount of iodine [11]. Iodine exists in the sea and the soil as iodide. Iodide ions are oxidised by sunlight to elemental iodine which is volatile. The iodine cycle in nature is complete if the concentration of iodide in the seawater is about 50–60 μg/L, approximately 0.7 μg/m^3 in the air, and the iodine in the atmosphere is returned to the soil through rain, with concentrations in the range of 1.8–8.5 μg/L. Iodine deficiency occurs in the soil when the return of the iodine to the soil is slow and in small amount compared to the original loss of iodine. Hence, all crops grown in this soil will be iodine deficient [12]. Low levels of iodine in the diet for people who do not get enough iodine from their food may lead to health problems collectively referred to as iodine deficiency disorders (IDD) [14].

2.2 The problem of IDD to human population

IDD is a major public health problem for population throughout the world which affects human from early foetal life through to adulthood [15]. Although IDD can affect any person of any age, pregnant women and children are the most vulnerable high-risk group for IDD [16]. Iodine requirement is high during pregnancy; it may increase by 50% because of increased maternal thyroxine production [14].

IDD in the foetus is the result of IDD in the mother, and this condition is associated with greater incidence of stillbirths, abortions, congenital abnormalities, neurological cretinism and psychomotor defects (**Figure 3**). In neonate, apart from mortality, the continuing severe IDD may affect the brain and physical development. Low birth weight is normally associated with a higher rate of congenital anomalies, and there were also evidences on substantial fall in infant mortality with improved birth weight following the iodized oil injection. IDD in the child and adolescent is associated with juvenile hypothyroidism, impaired mental function and retarded physical development. Studies on schoolchildren living in iodine-deficient areas indicated impaired school performance and IQs [17], while IDD in the adult had effects on their individual capacity, initiative and decision-making (**Figure 3**). These results indicate that IDD can be a major obstacle to human and social development of population living in an iodine-deficient environment. Therefore, correction of iodine deficiency is considered as a major contribution to population development [18].

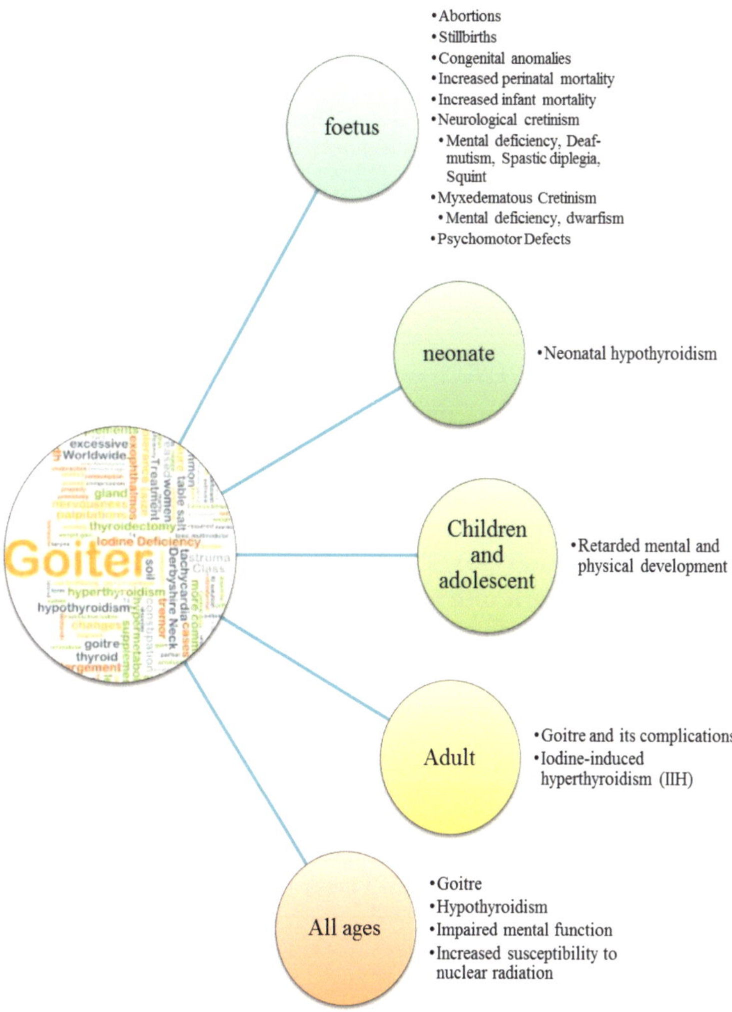

Figure 3.
Effect of iodine deficiency and the spectrum of IDD across the life span (clinical presentation) in various age groups, i.e. foetus, neonate, infant, child, adolescent and adult.

3. Iodine needs in pregnant women and their foetuses

Pregnant woman is one of the most susceptible groups for iodine deficiency. An adequate intake of iodine in the diet of pregnant woman is important to ensure normal growth and development of the foetus. During pregnancy, iodine requirement increases substantially to ensure adequate supply to the foetus particularly for healthy brain development. Iodine deficiency during pregnancy can cause maternal and foetal hypothyroidism and impairs neurological development of the foetus since it is secondary to transplacental passage of iodide. The consequences depend upon the timing and severity of hypothyroidism; the most severe manifestation is cretinism [14].

Since most foods have relatively low iodine content, Universal Salt Iodization (USI) programmes are executed. However, in countries or setting where USI is not possible, other mode of iodine supplementation needs to be implemented. It is difficult to accurately quantify iodine intakes using traditional methods of dietary assessment in terms of the contribution of iodized salt use from table and cooking to total iodine intake. In view of the limitations of dietary assessment for adequate iodine, a mUIC of 150–249 μg/L has been established to determine the adequate iodine status among pregnant women. However, the large intra-individual variation in UIC from either spot or 24-hour urine samples means that UIC cannot be used to assess iodine status in an individual pregnant woman. Therefore, the association between iodine status in pregnancy and the developmental outcome of the individual child is rather difficult to be assessed.

4. IDD elimination programme worldwide and Universal Salt Iodization (USI) programme

Universal Salt Iodization (USI) is currently the most widely used strategy towards sustainable control and elimination of IDD. There was a significant progress since the adoption of USI as a primary strategy to address IDD in 1993. Iodizing table salt is one of the best and least expensive methods of preventing IDD. Salt is used as a key vehicle as it is widely available and consumed in a regular amount throughout the year apart from a very low cost of salt iodization with US$ 0.05 per person per year [1]. This strategy has been implemented in most countries where iodine deficiency is a public health problem.

Various concerted global efforts have also been undertaken to eliminate IDD. This includes extensive advocacy from the international partners/alliances such as the WHO, UNICEF and ICCIDD (International Council for Control of Iodine Deficiency Disorders). These alliances have been in the forefront in helping countries to set up national salt iodization programmes. As reported by WHO/UNCEF/ICCIDD, there are currently 65 countries worldwide implementing USI as an effective strategy to eliminate IDD [3]. However, an effective USI in correcting iodine deficiency adequately through iodized salt must reach the whole affected population including pregnant women and children. Therefore, close and regular monitoring of iodized salt at various levels from the production, importation, retailer and household is crucial. Such monitoring requires close collaboration between the governments, salt industries and importers. In countries or areas within countries where USI is not possible, iodine supplementation needs to be implemented and especially targeted to pregnant women and children until USI is scaled up.

5. Urinary iodine results and their implications

Upon consumption, the needed amount of iodine is retained in the body, while excess iodine is excreted. Thus, high urinary iodine concentration does not reflect a disease state yet, but if persists on repeated urine sample, further blood tests is recommended. Urinary iodine reflects the food consumption taken overnight, and it is just an immediate biomarker (short-term reflection) for iodine intake. Thus, the long-term reflection of iodine intake will be more representative by measuring blood thyroglobulin (TG) as it is synthesised parallel to the amount of iodine present in the thyroid follicular cells. The very low or high individual urinary iodine readings will usually be repeated for testing to ensure that it is replicating the first reading.

6. Other blood biomarker measurement to support iodine-deficient status determined through urinary iodine measurement

For respondents with high mUIC, other biomarkers such as blood thyroglobulin (TG), thyroid-stimulating hormone (TSH), thyroid hormones (free thyroxine, fT_4;

	Whole group (n = 870)			Constraint group (n = 453)		
Hormone	Median	Minimum-Maximum	2.5th–97.5th percentiles[a]	Median	Minimum-Maximum	2.5th–97.5th percentiles[a]
TSH, mIU/L	1.31	0.05–14.50	0.30–3.63	1.36	0.12–5.29	0.40–3.77[b]
T_4, nmol/L	101.00	52.10–209.00	71.50–158.00	98.30	61.80–173.00	70.50–157.00
T_3, nmol/L	1.77	0.89–4.00	1.23–2.80	1.77	0.89–3.96	1.27–2.79
FT_4, pmol/L	16.20	8.24–28.60	12.70–20.80	16.20	9.30–24.70	12.80–20.40
FT_3, pmol/L	5.10	2.52–9.96	3.89–6.66	5.13	2.52–9.96	4.02–6.79

[a]*Data are reported as empirical percentiles.*
[b]*The lower limit of the TSH reference interval in the constraint group was >5% different from the comparable limit of the whole group.*

Table 2.
Reference intervals for thyroid hormones from subjects with normal thyroid gland assessed sonographically (table reproduced with permission of the rights holder, Clinical Chemistry) [19].

free triiodothyronine, fT_3) and thyroid antibodies, thyroid peroxidase antibodies (TPO Ab) and thyroglobulin antibodies (TgAb) are thus suggested for further diagnosis. Brief information on these biomarkers is as stated below:

6.1 Free thyroxine (fT_4) and free triiodothyronine (fT_3)

T_4 is a molecule of thyronine bond to four atoms of iodine, while T_3 has three iodine atoms. T_4 is more abundant than T_3, but through deiodinase activity, the more potent T_3 is synthesised. Low level of T_4 usually indicates hypothyroidism. For clinical biochemistry free T_4 (fT_4) and free T_3 (fT_3) are usually measured as these are the biologically active forms. Reference intervals for fT_4 and fT_3 are as stated in **Table 2**.

6.2 Thyroid-stimulating hormone (TSH)

TSH is released upon the induction by thyrotropin-releasing hormone (TRH), secreted from the hypothalamus. TRH is secreted when serum T_3 and/or T_4 is low. TSH induces the production of thyroid hormones T_3 and T_4. Low level of TSH usually indicates hyperthyroidism (parallel to high levels of T_3 and/or T_4) [17]. Reference interval for TSH is as stated in **Table 2**.

6.3 Thyroglobulin (TG)

TG is the globulin where binding of iodine to tyrosine to form thyroid hormones takes place. It is the long-term biomarker for iodine status in a human, besides urinary iodine as the short-term biomarker for iodine nutrition.

The median for reference interval for dried blood spot (DBS) TG from 5- to 14-year-old children before intervention of iodized salt is 49 g/L. After using iodized salt for 5 months, the DBS-TG decreased to 13 g/L and further decreases to 8 g/L after 10 months of consumption of iodized salt [20].

6.4 Thyroid antibodies [thyroid peroxidase antibodies (TPO Ab) and thyroglobulin antibodies (TgAb)]

This antibody is for binding to the antigen thyroid peroxidase (TPO) enzyme which is responsible for thyroid hormones synthesis. Once the enzyme is bound to the antibody, less free unbound enzymes are available for thyroid hormone

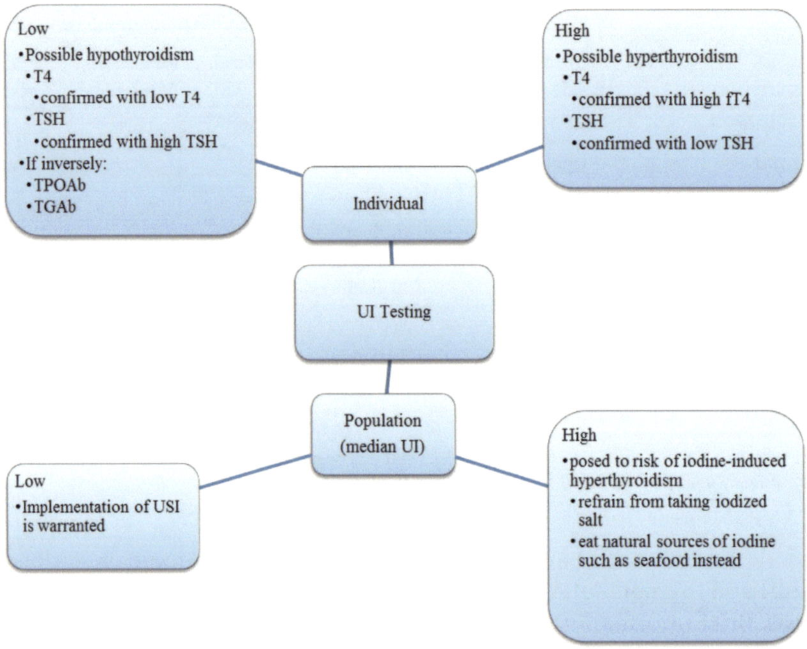

Figure 4.
Subsequent biomarker testing and further action for individual and population urinary iodine estimation.

synthesis, thus causing lower thyroid hormone production. TgAb is the antibody for the globulin TG. With the presence of elevated amount of TgAb, the thyroglobulin (TG) will bind to its antibody, and lesser TG is available to bind to the thyroid hormones for transportation in the blood vessels, thus causing lesser thyroid hormones being circulated in the human body [21]. The reference intervals are 15 kIU/L for TPOAb and 31 kIU/L for TgAb [22] (**Figure 4**).

7. Diagnosis for the iodine-induced thyroid dysfunction in respondents

Both insufficient and excessive iodine intake can result in thyroid dysfunctional diseases. If the thyroid diseases are due to iodine-induced phenomena, the main management is to avoid or reduce iodine intake, followed by the appropriate drugs if symptomatic or there is abnormality with the thyroid function test (TFT) results.

7.1 Iodine-induced hypothyroidism

Iodine-induced hypothyroidism can occur in normal individuals and in those with chronic systemic disease and underlying thyroid disorders. It has been seen in patients who had a history of post-partum thyroiditis and subacute thyroiditis and in those treated with recombinant interferon-alpha. The hypothyroidism was described as transient, and thyroid function returns to normal in 2–3 weeks after iodide withdrawal. Some patients may require transient T_4 replacement therapy [23].

7.1.1 Clinical presentation

The presenting clinical features of hypothyroidism depend on the duration and severity, the nature of its onset and the patient's psychological characteristics [24]. The following are the signs and symptoms of hypothyroidism:

- Fatigue
- Weight gain from fluid retention
- Dry skin and cold intolerance
- Yellow skin
- Coarseness or loss of hair
- Hoarseness
- Goitre
- Reflex delay, relaxation phase
- Ataxia
- Constipation
- Memory and mental impairment
- Decreased concentration
- Depression
- Irregular or heavy menses and infertility
- Myalgias
- Hyperlipidaemia
- Bradycardia and hypothermia
- Myxoedema fluid infiltration of tissues

7.1.2 Diagnosis

When there is clinical suspicion of hypothyroidism, a thyroid function test should be performed. Measurement of TSH level is the primary test to confirm primary hypothyroidism. Other laboratory evaluation may include free T_4 and thyroid antibodies (anti-thyroid peroxidase and anti-thyroglobulin autoantibodies). Imaging studies to evaluate any structural thyroid abnormalities include a thyroid scan, ultrasonography or both [24]. When iodine deficiency occurs during pregnancy, it is associated with foetal hypothyroidism, mental impairment and increased neonatal and infant mortality [25]. In adults, iodine-induced hypothyroidism is rare. The most common manifestation is goitre. Low iodine intake leads to reduced T_4 and T_3 production which results in increased TSH secretion in an attempt to restore normal T_4 and T_3 production. TSH also stimulates thyroid growth leading to goitre. The goitre is initially diffuse but progresses to nodular goitre and eventually to thyroid autonomy and possible hyperthyroidism [26]. Excess iodine ingestion or exposure above the limit of the recommended daily iodine intake induces thyroid dysfunction. Iodine-induced thyroid dysfunction may be subclinical or overt. Excess iodine is generally well tolerated. However, individuals with underlying thyroid disease or other risk factors may be susceptible to iodine-induced thyroid dysfunction following acute or chronic exposure.

7.1.2.1. Predisposing risk factors in iodine-induced hypothyroidism

Individuals with underlying thyroid disease:

a. Euthyroid Graves' disease previously treated by radioactive iodine, thyroidectomy or anti-thyroid drugs

b. Hashimoto's thyroiditis

c. Euthyroid with a history of subacute thyroiditis

d. Euthyroid with a history of post-partum thyroiditis

e. Euthyroid with a history of type 2 amiodarone-induced thyrotoxicosis

f. Euthyroid posthemithyroidectomy

g. Euthyroid after interferon-alpha therapy

The spectrum of iodine deficiency disorders (IDD) is seen across the life span in various age groups, i.e. foetus, neonate, infants, child, adolescent and adult. They include endemic goitre and cretinism, endemic mental retardation, decreased fertility rate, increased perinatal death and infant mortality and varying degrees of other growth and developmental abnormalities (**Table 3**) [27, 28]. Hypothyroidism due to very low iodine intake is now extremely rare. Adults usually have the typical clinical manifestations of hypothyroidism and goitres [29].

Life stage	Spectrum of IDD
Foetus	Abortions Deaf mutism Stillbirths Congenital anomalies Increased perinatal mortality Endemic cretinism
Neonate	Neonatal goitre Neonatal hypothyroidism Endemic mental retardation Increased susceptibility of the thyroid gland to nuclear radiation
Child and adolescent	Goitre Subclinical hypothyroidism Impaired mental retardation Retarded physical development Increased susceptibility of the thyroid gland to nuclear radiation
Adult	Goitre with its complications Hypothyroidism Impaired mental function Hyperthyroidism in the elderly (after iodized salt)

Source: Refs. [27, 28].

Table 3.
The spectrum of IDD across the life span.

7.2 Iodine-induced hyperthyroidism

Iodine-induced hyperthyroidism can occur after intake of excess iodine in the diet, exposure to radiographic contrast media for imaging procedures or medications [30]. In iodine-sufficient areas, iodine can induce hyperthyroidism in euthyroid patients with previous thyroid diseases. These include patients who were treated with anti-thyroid drugs for Grave's disease and post-partum thyroiditis [31, 32].

7.2.1 Clinical presentation

The severity and spectrum of symptoms and signs of hyperthyroidism may be related to the duration of the illness, the effects of excess thyroid hormone and the age of the patient [24].

The symptoms and signs include the following:

- Nervousness and irritability
- Palpitations and tachycardia
- Heat intolerance or increased sweating
- Tremor
- Weight loss or gain
- Alterations in appetite
- Frequent bowel movements or diarrhoea
- Dependent lower-extremity oedema
- Sudden paralysis
- Exertional intolerance and dyspnoea
- Menstrual disturbance (decrease flow)
- Impaired fertility
- Mental disturbances
- Sleep disturbances (including insomnia)
- Changes in vision, photophobia, eye irritation, diplopia or exophthalmos
- Fatigue and muscle weakness
- Goitre (depending on cause)
- Pretibial myxoedema

8. Novel strategies for the vulnerable group

In the Iodine Global Network newsletter published on its website, Prof. Dr. Zimmermann had laid out the strategies needed for the vulnerable groups, i.e., the newborns, infants and the children. He had suggested that these tests should be done on these groups respectively: blood thyroid-stimulating hormone, urinary iodine and blood spot thyroglobulin (http://www.ign.org/zimmermann-calls-for-new-strategies-against-idd.htm). He also suggested using the TSH measurement in the newborn screening for iodine assessment. Commonly, iodine-deficient newborns present with elevated TSH. The WHO reported that if the TSH level is greater than 5 mIU/L from the whole blood of 3% of the newborns measured after 3–4 days post-birth, this would indicate that the population is iodine deficient. Another suggestion made by him was to use the newborns' urinary iodine as a marker of IDD in addition to the current practise of measuring the median urinary iodine of the schoolchildren. Using a non-invasive system, urine was collected from infants, and study had shown that in 1200 infants, the baseline TSH of 77 μg/L had increased to 100 μg/L after 4 days post-birth. In addition to that, a system for collecting young children blood spot was done to measure their TG concentration, and a reference range of 4–40 μg/L had been determined. Noteworthy, there are challenges in order to establish a specific international reference range among newborns as there will be differences between the population from the USI areas and the non-USI areas.

9. Conclusion

Excess urinary iodine is generally well tolerated, but individuals with underlying thyroid disease or other risk factors may be susceptible to iodine-induced thyroid dysfunction following acute or chronic exposure. Increased iodine exposure including the global public health efforts of iodine supplementation, the escalating use of iodinated contrast radiologic studies, amiodarone administration in vulnerable patients, excess seaweed consumption and various miscellaneous sources should be looked for.

Iodine-induced thyroid dysfunction may be subclinical or overt. Recognition of the association between iodine excess and iodine-induced hypothyroidism or hyperthyroidism is important in the differential diagnosis of patients who present without a known cause of thyroid dysfunction.

Conflict of interest

It is declared that there is no conflict of interest involved in the publication of this book chapter.

Author details

Husniza Hussain[1*], Rusidah Selamat[2], Lim Kuang Kuay[3], Fuziah Md Zain[4] and Muhammad Yazid Jalaludin[5]

1 Nutrition Unit, Cardiovascular, Diabetes and Nutrition Research Centre, Institute for Medical Research, Ministry of Health Malaysia, Kuala Lumpur, Malaysia

2 Nutrition Division, Ministry of Health Malaysia, Putrajaya, Malaysia

3 Centre for Occupational Health Research, Institute for Public Health, Ministry of Health Malaysia, Kuala Lumpur, Malaysia

4 Paediatrics Department, Putrajaya Hospital, Ministry of Health Malaysia, Putrajaya, Malaysia

5 Faculty of Medicine, Department of Paediatrics, University of Malaya, Kuala Lumpur, Malaysia

*Address all correspondence to: husnizah9thesis@gmail.com

References

[1] Zimmermann MB. Iodine requirements and the risks and benefits of correcting iodine deficiency in populations. Journal of Trace Elements in Medicine and Biology. 2008;**22**:81-92

[2] Jooste PL, Strydom E. Methods for determination of iodine in urine and salt. Best Practice & Research Clinical Endocrinology & Metabolism. 2010;**24**(1):77-88

[3] WHO/UNICEF/ICCIDD. Assessment of Iodine Deficiency Disorders and Monitoring Their Elimination. 3rd ed. France: World Health Organization; 2007. pp. 7-10

[4] Aburto NJ, Abudou M, Candeias V, Wu T. Effect and Safety of Salt Iodization to Prevent Iodine Deficiency Disorders: A Systematic Review with Meta-Analyses. Geneva: World Health Organization; 2014

[5] Danijela RM, Zlata P, Lee H, Jiri R, Amelie C, Kate A, et al. Methods of assessment of iodine status in humans: A systematic review. The American Journal of Clinical Nutrition. 2009;**89**(suppl):2052S-2069S

[6] Pincock S. Basil Hetzel: vanquishing iodine deficiency disorders. The Lancet. 2013;**381**(9868):717

[7] Kopp P. Thyroid Hormone Synthesis. Werner and Ingbar's the Thyroid: A Fundamental and Clinical Text. 10th ed. Philadelphia: Lippincott Williams & Wilkins; 2012. pp. 48-74

[8] Farhana A, Shaiq AG. Iodine, iodine metabolism and iodine deficiency disorders revisited. The Indian Journal of Endocrinology and Metabolism. 2010;**14**(1):13-17

[9] American Thyroid Association. Hypothyroidism: A Booklet for Patient and Their Families. Falls Church: American Thyroid Association; 2013. pp. 598-606

[10] Clugston GA, Hetzel BS. Iodine. In: Shils ME, Olson JA, Shike M, editors. Modern Nutrition in Health and Disease. 11th ed. Philadelphia: Lea & Febiger; 2014

[11] Speeckaert MM, Speeckaert R, Wierckx K, Delanghe JR, Kaufman JM. Value and pitfalls in iodine fortification and supplementation in the 21st century. The British Journal of Nutrition. 2011;**106**:964-973

[12] Linhares DPS, Garcia PV, Almada A, Ferreira T, Queiroz G, Cruz JV, et al. Iodine environmental availability and human intake in oceanis island: Azores as a case study. Science of the Total Environment. 2015;**538**:531-538

[13] Balucan FS, Morshed SA, Davies TF. Thyroid autoantibodies in pregnancy: Their role, regulation and clinical relevance. Journal of Thyroid Research. 2013;**2013**(182472):15. http://dx.doi.org/10.1155/2013/182472

[14] Zimmermann MB. The role of iodine in human growth and development. Seminars in Cell & Developmental Biology. 2011;**22**:645-652

[15] Hetzel B. Vanquishing iodine deficiency disorders. Lancet. 2013;**381**(9868):717. Available from: http://www.ign.org/zimmermann-calls-for-new-strategies-against-idd.htm [Accessed: 19 November, 2018]

[16] Trumpff C, Schepper JD, Tafforeau OHV, vanderfaeillie J, Vandevijvere S. Mild iodine deficiency in pregnancy in Europe and its consequences for cognitive and psychomotor development of children: A review.

Journal of Trace Elements in Medicine and Biology. 2013;**27**:174-183

[17] Costeira MJ, Oliveira P, Santos NC, Ares S, Saenz-Rico B, de Escobar GM, et al. Psychomotor development of children from an iodine deficient region. The Journal of Pediatrics. 2011;**159**(3):447-453

[18] UNICEF. Sustainable of Iodine Deficiency. New York: The United Nations of Children's Fund; 2008. pp. 5-7

[19] Kratzsch J, Fiedler GM, Leichtle A, Brügel M, Buchbinder S, Otto L, et al. New reference intervals for thyrotropin and thyroid hormones based on National Academy of Clinical Biochemistry criteria and regular ultrasonography of the thyroid. Clinical Chemistry. Aug 2005;**51**(8):1480-1486. DOI: 10.1373/clinchem.2004.047399

[20] Zimmermann MB, de Benoist B, Corigliano S, Jooste PL, Molinari L, Moosa K, et al. Assessment of iodine status using dried blood spot thyroglobulin: Development of reference material and establishment of an international reference range in iodine-sufficient children. The Journal of Clinical Endocrinology and Metabolism. 2006;**91**(12):4881-4887

[21] Spencer CA, Takeuchi M, Kazarosyan M, Wang CC, Guttler RB, Singer PA, et al. Serum thyroglobulin autoantibodies: Prevalence, influence on serum thyroglobulin measurement, and prognostic significance in patients with differentiated thyroid carcinoma. The Journal of Clinical Endocrinology and Metabolism. 1998;**83**(4):1121-1127

[22] Jensen EA, Petersen PH, Blaabjerg O, Hansen PS, Brix TH, Hegedüs L. Establishment of reference distributions and decision values for thyroid antibodies against thyroid peroxidase (TPOAb), thyroglobulin (TgAb) and the thyrotropin receptor (TRAb). Clinical Chemistry and Laboratory Medicine (CCLM). 2006;**44**(8):991-998

[23] Markou K, Georgopoulos N, Kyriazopoulou V, Vagenakis AG. Iodine-induced hypothyroidism. Thyroid. 2001;**11**(5):501-510

[24] American Association of Clinical Endocrinologists. Medical Guidelines for Clinical Practice for the evaluation and treatment of hyperthyroidism and hypothyroidism. AACE Thyroid Task Force. Endocrine Practice. 2002;**8**(6):457-469

[25] Glinoer D. The importance of iodine nutrition during pregnancy. Public Health Nutrition. 2007;**10**(12A):1542-1546

[26] Krohn K, Paschke R. Progress in understanding the etiology of thyroid autonomy. The Journal of Clinical Endocrinology and Metabolism. 2001;**86**(7):3336-3345

[27] Knobel M, Medeiros-Neto G. Pediatric aspects of thyroid function and iodine. In: Diseases of the Thyroid in Childhood and Adolescence. Vol. 11. Karger Publishers; 2007. pp. 56-79

[28] Knobel M, Medeiros-Neto G. Relevance of iodine intake as a reputed predisposing factor for thyroid cancer. Arquivos Brasileiros de Endocrinologia e Metabologia. 2007;**51**(5):701-712

[29] Garber JR. Woeber for the American Association of Clinical Endocrinologists and American Thyroid Association Taskforce on hypothyroidism in adults KA. Clinical practice guidelines for hypothyroidism in adults: Cosponsored by the American Association of Clinical Endocrinologists and the American Thyroid Association. Thyroid. 2012;**22**:1200-1235

[30] Reid JR, Wheeler SF. Hyperthyroidism: Diagnosis and treatment. American Family Physician. 2005;**72**(4):623-630

[31] Topliss DJ, Eastman CJ. Diagnosis and management of hyperthyroidism and hypothyroidism. The Medical Journal of Australia. 2004;**180**(4):186-194

[32] Zimmermann MB, Jooste PL, Pandav CS. Iodine-deficiency disorders. Lancet. 2008;**372**:1251-1262

Chapter 3

Advancement in Analytical and Bioanalytical Techniques as a Boon to Medical Sciences

Khushaboo Pandey and Om Prakash Mishra

Abstract

The most important objectives frequently found in analytical and bioanalytical chemistry involve advancement of analytical techniques and its application to relevant medical/clinical problems. Keeping in view to these aspects, the present chapter is primarily focused on the development of advanced analytical techniques applied in the medical field. For example, N-acetyl-beta-D-glucosaminidase (NAG) enzyme is a specific biomarker of acute kidney injury. A biomarker is an entity that is purposely measured and estimated as an indicator of normal biological process, pathogenic process, or pharmacological responses to a therapeutic intervention. Hence, successive measurements of urinary NAG may enhance its clinical use as an indicator of ongoing tubular injury. Hence, in order to obtain information for selective monitoring of biomarker, the development of a practical and valid analytical method is important. Experimentation is driven by the need to know more about the medical effects and safety features of the biologically active analyte. It is therefore more important to evaluate the information that is already available for that particular analyte and to quantify the level of uncertainty for the proposed technique.

Keywords: analytical, bioanalytical, biomarker, clinical, medical, N-acetyl-beta-D-glucosaminidase

1. Introduction

This chapter represents a collective attempt to present a wide range of analytical techniques applied for the clinical development process. Analytical chemistry is concerned with the chemical characterization of matter and refining the qualitative and quantitative problem about that matter. It plays a vital role in almost all the aspects of scientific research and development, for example, clinical, forensic, environmental, and pharmaceutical sciences. In medicine, analytical chemistry is the key for clinical laboratory tests which imparts basis of disease diagnosis and chart progress for recovery to the physicians [1]. **Figure 1** describes the scheme through which physicians ruled out or analyze the disease prognosis and therapeutic drug monitoring in patients. In accord with this, an analytical chemist also explores the idea of developing advanced technique for betterment of human healthcare and in sorting out the problems related to the disease diagnosis. Implementation of an analytical technique mainly depends on the varying degree of selectivity, sensitivity, accuracy, precision, cost, and rapidity of that particular technique. The techniques

Figure 1.
Schematic representation of four common decision making steps in which the result of an investigation is involved.

employed may be based either on physical property or chemical property of an analyte. An analyte is defined as a constituent which has to be determined in a given sample type. The classical analytical techniques include gravimetric, volumetric, and titrimetric methods; on the other hand, instrumental techniques involve ultraviolet-visible (UV-Vis), infrared (IR), and near-infrared (NIR) spectrophotometry fluorimetry, atomic spectroscopy (absorption/emission), electroanalytical chromatography, and radioimmunoassay. Instrumental techniques are usually more sensitive and selective than classical techniques but are less precise. Precision of techniques means the repeatability of a result and is expressed as standard deviation. Selectivity of an analytical method defines the measurement of a particular analyte from sample solution to a certain degree, in the presence of other analytes, without any interference. However, sensitivity of a method describes the ability to recognize two different concentrations.

However, medical and clinical analyses are undergoing the greatest extension of instrumental methods [2]. Interest in identifying biologically active compounds is growing rapidly and providing new challenges for the analytical chemists. These challenges have been resolved by the introduction of bioanalytical technique as a modern approach to disease diagnosis and therapy. A bioanalytical method is a combination of different procedures which are (i) collection, (ii) processing, (iii) storage, and (iv) analysis of a biological sample (blood-cerebrospinal fluid (CSF), serum, plasma, or urine, tissue, and skin). This method is also useful for quantitative determination of drugs and metabolites in biological samples. For that reason, technologies used to perform bioanalytical methods vary according to the analyte's nature. Hence, to find out the appropriate technologies involved in a bioanalytical method for the purpose of quantification of an analyte, the method validation is important. This procedure is termed as bioanalytical method validation (BMV). Few techniques commonly applied in bioanalytical studies include hyphenated (combination of two techniques) techniques like liquid chromatography (LC), gas chromatography (GC), capillary electrophoresis (CE) coupled with mass spectrometry (MS), and advanced automated chromatographic techniques, for example, high-performance liquid chromatography (HPLC) [3].

During the past decades, the analytes that have been targeted in bioanalytical studies include amino acid, peptides, proteins, serum enzyme, tumor and cancer genes, carbohydrates, vitamins, catecholamines, cardiac risk factors, etc. [4].

With the recurrent analysis of biomolecules, numerous analytical techniques and instrumentation have been evolved and applied in the field of medical sciences which are as follows:

- Sensors
- Electrophoresis
- Chromatography
- Mass spectrometry
- Optical techniques (microscopy)
- Radio- and immunochemical techniques
- Hyphenated techniques
- Point-of-care instrumentation

Biological samples have the potential to deliver important biomarkers in the clinic due to accessibility of these biological materials. In clinical development, the most important benefit offered by biomarkers is to limit investigational drugs to critical care patients who would gain the therapies to observe the effectiveness of those drugs [5]. The role of a biomarker is to give information about the biological mechanism involved within a disease or treatment of disease having the capability to correlate with the clinical findings. One of the most tangible problems that research scientists are facing in recent years is finding disease biomarkers that are translated well from animal or computer simulation to humans. For example, increase in enzyme activity in computer or animal model may have a significant impact in theoretical computer or animal model, whereas same enzyme activity enhancement may have a very limited or no clinical impact.

There is no denying that "analytical and bioanalytical technique" is a broad topic, incorporating technologies from classical chromatography to point-of-care instrumentation. But unifying and doing them as quickly, accurately, and inexpensively as possible are drives to make chemical or biochemical measurements. Over the preceding sections, we would study the technological improvements along those lines across the broad field of analytical methods. Every subsection of analytical techniques applied in medical field has experienced improvement and advancement as well.

Researchers are interested in mapping the neural connectivity of the brain through scanning electron microscopy (SEM). This could be now employed with more powerful microscopes, such as focused ion beam and multi-beam SEM, to collect serial images of ultrathin brain slices [6]. They can now build surface plasmon resonance substrates out of silver rather than the more typical gold and an SPR microscope to image and quantify 1296 binding events in parallel [7]. Those scientists who are interested in surface properties can now scan those surfaces faster than ever, thanks to high-speed atomic force microscopy [8].

2. Sample collection and storage

We have to keep in mind that biological samples, collected from the patients, must be transported to the initial assessment center as soon as possible.

The type of preservatives should be known to protect the samples from degradation prior to cryopreservation at a reasonable cost. Cryopreservation is a process to store biological samples at very low temperature for prevention of damage. The purpose is to find readily accessible and data-rich biological samples. The stability of a wide range of bioanalytes and cells as a component of whole blood should be estimated, taking into account different anticoagulant (inhibition of coagulation of blood) media, at different temperatures and under varying transport conditions. Bioanalytes can be known biochemicals, such as DNA, defined proteins, and specific metabolites, or unknown analytes, such as the constituent plasma/serum proteome and metabonome [9].

Design and testing of the sample handling protocol considered as key factors that affect the stability of biological samples, including anticoagulants, stabilizing agents, and temperature, elapsed time from collection to initial processing and endogenous degrading properties (enzymes, cell death). We also aim for cost-efficiency by avoiding collecting multiple sources of material for the same analyte. The samples undergo minimal processing locally in the assessment centers before being shipped to the main laboratory for processing with the aim of cryopreservation within 24 h of collection. Samples are protected against degradation during shipping by being chilled at 4°C (only peripheral blood lymphocytes, at 18°C). Once the samples get processed in the laboratory, they are placed in cabinets maintained at −80°C for the working archive or in nitrogen vapor at −180°C or below for the backup archive.

3. Sample preparation

Biological samples involve plasma, serum, CSF, bile, urine, tissue homogenates, saliva, seminal fluid, and frequently whole blood. Quantitative analysis of drugs

Sample preparation techniques	Advantages
Liquid phase extraction (LLE)	LLE is one of the first methods used for extraction. It depends on the partitioning of analytes between two immiscible liquids. The resulting extract may be directly analyzed or further purified and concentrated by subsequent LLE and evaporations.
Solid phase extraction (SPE)	SPE is a method for the isolation and concentration of selected analytes from a fluid sample by their transfer on a solid phase. The analytes are recovered by elution or thermal desorption. This method has high recovery, uses less organic solvent than LLE.
Affinity separation: molecularly imprinted polymers (MIPs)/ antibodies	The affinity sorbent may consist of an immobilized antibody or a molecularly imprinted polymer. This technique is highly specific and very sensitive, but the sorbent is difficult to prepare; it suffers from cross-reactivity and leaking of template.
Solid phase micro extraction (SPME)	SPME, a solvent free extraction method, consists of a single extraction step, but the experimental variables must be well controlled. It reduces solvent and sample volume needs and sample preparation time. Improves detection limits i.e., parts per trillion level detection.
Ultrafiltration and microdialysis (MD)	Ultrafiltration consists of filtering the sample through a special size-excluding filter, either by applying pressure (10–100 psi) or by centrifugation. The method is widely used, simple, efficient, but suffers from ligand binding to the filter and shift of equilibrium. Dialysis and MD can be used to separate an analyte by diffusion through a semi-permeable membrane.

Table 1.
Sample preparation techniques used for biological samples [32].

and metabolites containing huge amounts of proteins and large numbers of endogenous compounds within these samples is very complicated. Direct injection of drug containing biological sample into a chromatographic column results in the precipitation or absorption of proteins on the column packing material, resulting in an immediate loss of column performance. A number of advances have made to convert sample preparation techniques, used for the cleanup of drugs in biological samples into formats that are acceptable for high-volume processing with or without automation. The most widely used cleanup methods for separation of biomolecules from biological samples are summarized in **Table 1**.

Because of this, sample preparation became a prominent step in the analysis of biological samples. In recent years, the necessity of new developed method is largely required. Frequently, it was earlier considered as a separate procedure prior to the analysis, while it nowadays has become a more or less integrated part of the analytical procedure. It is necessary to lay the foundation of their development on a systematic and scientific approach. Thus, fundamental understanding of the different processes involved in a sample preparation method is served as a basis for its optimization [10].

We should select the appropriate sample preparation method on the basis of requirements of the assay and time allowed to run sample preparation method.

4. Spectrophotometry

The spectrophotometric technique is used to study interactions between electromagnetic radiations and analyte (**Figure 2**). The concentration of an analyte is determined by using a graph which is called standard analytical curve. An example is determination of iron in blood serum. The iron content of blood serum is determined after deprotonation (by precipitating protein) with trichloroacetic acid and reduction with hydroxyl ammonium sulfate. Iron (II) ions are reacted in the medium buffered with ammonium acetate and with diphenyl-1,10-phenanthroline-disulfonic acid disodium salt (bathophenanthroline disulfonate–Na), and the absorbance of the complex formed is measured at 535 nanometer. The concentration belonging to the absorbance data of the test solution is read from the standard analytical curve and multiplied by three for threefold dilution [11].

4.1 Optical spectroscopic techniques of human models

Optical spectroscopy for biomedical applications covers up the plethora of medical technological and fundamental research areas. This includes screening and early detection of diseases which remain clinically silent over long periods. It is a noninvasive, fast spectroscopic technique. The technique is also capable of observations from femtosecond time scale at nanometer spatial resolution, so it can be applied in all areas of life sciences. This technique is to make an early, noninvasive, and patient-specific diagnosis near the source of the disease and then to treat the

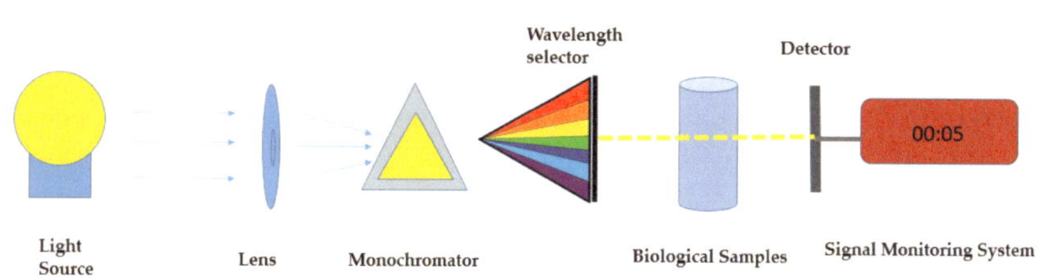

Figure 2.
Schematic representation of biological sample determination using spectrophotometric technique.

disease at primary stage, for example, Alzheimer's disease and coronary disease. Thus, optics offers a wide variety of diagnostic methods and products of biomedical spectroscopy [12].

4.2 Absorptions and reflectance spectroscopy

This method involves investigations of brain dysfunction and mental health problems like depression, epilepsy, and Alzheimer's disease [13]. The direct absorption like near-infrared (NIR) techniques and instrumentation are particularly suitable in routine neonatal care applications.

However, diffused reflectance spectroscopy, in the UV-Vis-NIR region, can be used for biomedical applications, like studies on skin condition (vitiligo, psoriasis, skin cancer) and glucose concentration measurement [14].

4.3 Photoacoustic spectroscopy (PAS)

This includes measurement of concentration of biomolecular species. Examples are glucose determination, characterization of tissue status (biopsy tissue), and imaging application. PAS can provide information on three-dimensional distribution of specific molecular species in a specimen, by appropriate choice of excitation wavelength [15].

4.4 Raman and infrared spectroscopy

The "fingerprint" molecular specific technique will be of great advantage in understanding the biochemical interactions involved in induction, progression, therapeutic invention, and regression. This technique is suitable for biomedical applications such as breath analysis, drug-cell interaction, microscopy, and imaging of biopsy sample (tissue, fine needle aspiration) [16–18].

4.5 Fluorescence spectroscopy

Depending on excitation wavelength, the fluorescence peak has been observed in blood and urine spectra. A few examples are spectra observed from epithelial tissues, proteins, NAH, FAD, and hemoglobin [19].

4.6 Mass spectrometry

Mass spectroscopy (MS) measures masses within the sample. In mass spectroscopy, chemical species get ionized and ions get sorted on the basis of their mass-to-charge ratio (**Figure 3**). Major application of MS includes confirmation of immunoassay-positive drug screens, identification of inborn errors of metabolism, and analysis of steroid hormones [20]. Conclusive identification of molecules that range in size from tens of daltons (small molecules) to hundreds of thousands of daltons (biomolecules) is based on different principles.

Discovery of highly sensitive polymerase chain reaction (PCR) was a major step forward in the biomedical research and diagnostics. This technique is used for analysis of small quantities of short sequences of DNA and RNA without cloning. PCR can detect the presence of pathogens earlier than the culture tests. The miniaturization of MS systems allows a transportable device that minimizes the need of highly skilled operators and allows for rapid and accurate MS analysis in a point-of-care format (near the physician's clinic) [21].

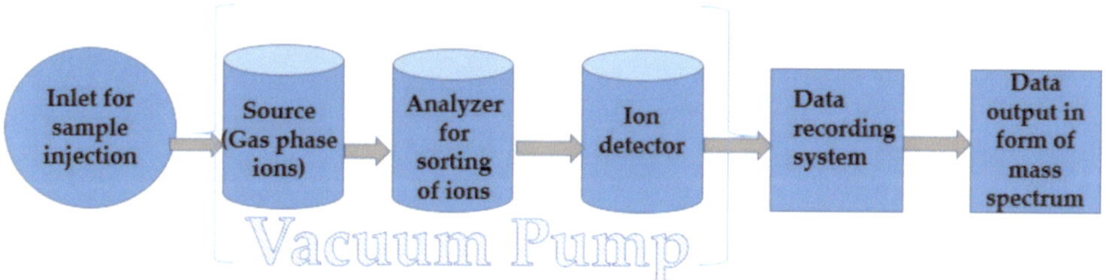

Figure 3.
Schematic representation of mass spectrometry detection of sample.

4.7 Imaging techniques

Advances in medical imaging present a great opportunity in drug development. A number of different imaging technologies are available. These include computed tomography (CT), magnetic resonance imaging (MRI), magnetic resonance spectroscopy (MRS), positron-emission tomography (PET), and single photon emission computed tomography (SPECT). If adequately qualified, imaging biomarkers can be very helpful in the early stages of clinical development [22].

5. Sensors

An electrochemical sensor consists of a diffusion barrier, a sensing electrode (working electrode, measuring electrode, or anode), a counter electrode (cathode), and an electrolyte. Their fabrication includes various types of systems such as conductometry, voltammetry, potentiometry, and capacitance, and it is an important tool to detect various analytes in environmental, clinical, and biological fields due to their high sensitivity, cheapness, and miniaturization. These sensors have the potential to achieve sensitive, specific, and low-cost detection of biomolecules which is relevant to the diagnosis and monitored treatment of disease [23]. A few are listed below:

5.1 Potentiometric sensors

Techniques based on measurement of potential sensor are termed as potentiometry, for example, determination of potassium in blood serum by direct potentiometry with an ion-selective electrode. The determination of urea is a frequent task of clinical laboratories. The basis of enzyme electrode function is the selective recognition of urea by urease enzyme. The following reaction takes place:

$$\text{Urea} + 3H_2O \xrightarrow{\text{Urease}} 2NH_4^+ + HCO_3^- + OH^- \quad (1)$$

This reaction can be followed using different potentiometric electrodes [24].

5.2 Molecularly imprinted polymer (MIP) sensors

MIP could be one of the important tailor-made systems for targeted analyte recognition exclusively even their presence in complex real biological samples in parts per million to parts per billion levels. The general idea is to create the cavity in the presence of the guest. The guest organizes and promotes energy-minimized interactions with polymer forming around it. Thus, after washing the guest out, the polymers retain

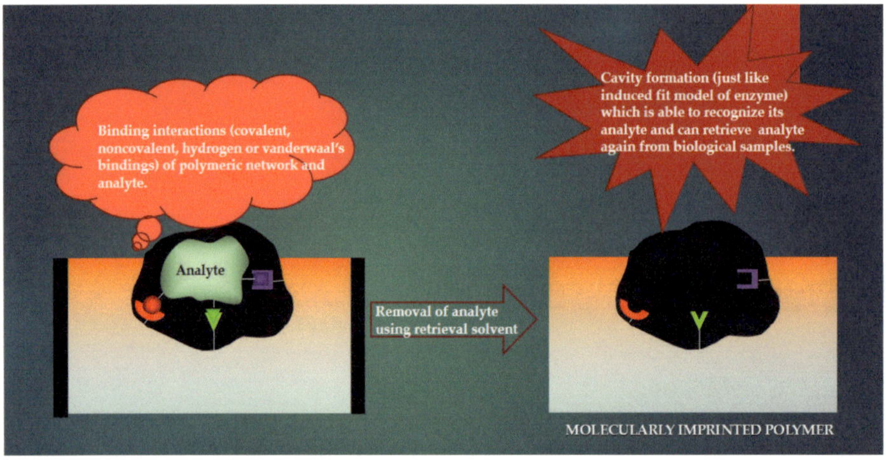

Figure 4.
Synthesis of molecularly imprinted polymer.

a template cavity of the guest's size and shape which subsequently display binding selectivity toward the guest just like induced-fit model of enzyme (**Figure 4**). The MIP-modified sensors can be used for biological and pharmaceutical analyte determination from biological samples. An example is the development of a polyscopoletin-based MIP nanofilm for the electrochemical determination of elevated human serum albumin (HSA) in urine. The results suggest that MIP-based sensors may be applicable for quantifying high-abundance proteins in a clinical setting [25].

5.3 Biosensors

The need for rapid, simple handheld testing devices in medicine paves the way for introduction of biosensor. Biological sensors are optical, electrical, and piezoelectrical devices that have the ability to detect biological compounds, such as nucleic acids and proteins [26, 27]. Early diagnosis of inherited disease is important for effective treatment and is sometimes lifesaving. Methods, like enzyme-linked immunosorbent assay and PCR, can require highly skilled professionals and expensive chemicals and can be time-consuming. In this area, many biosensor schemes had developed as an alternative to classical methods.

The latest advancements in nanotechnology result in its application for cancer biomarker recognition [28]. Several other biosensors include nanomaterial-based biosensors [29], peptide nucleic acid-based biosensors, biosensors for medical mycology, optical DNA biosensors, and last but not least, the biosensors for the diagnosis of heart disease [30].

6. Chromatographic separation techniques

In chromatographic separation technique, various constituents of the mixture in the given sample travel at different speeds, causing them to separate. This technique involves two phases: a mobile phase and a stationary phase. The separation mainly depends on the differential partitioning between these two phases. A bonded phase is a stationary phase that is covalently bonded to the support particles or to the inside wall of the column tubing. The stationary phase is the substance fixed in place for the chromatography procedure, and the mobile phase is moving in a definite direction [31]. Examples include the silica layer in thin layer chromatography (TLC). Archer John Porter Martin and Richard Laurence Millington Synge won a Nobel Prize in chemistry for chromatography invention [32]. Their work encouraged the rapid development of several advanced chromatographic methods such as paper chromatography, gas chromatography, and HPLC. The differences in a compound's partition coefficient bring about differential retention on the stationary phase and thus affect the separation process. Chromatography may be preparative or analytical. The preparative chromatography separates the components of a mixture for later use and is thus a form of purification. Analytical chromatography is done normally with smaller amounts of material and is for establishing the presence or measuring the relative proportions of analytes in a mixture. **Figure 5** describes a chromatogram for a biological system where the signal is proportional to the concentration of the specific analyte separated.

Depending upon the shape of stationary phase, chromatography may be (i) planar chromatography, having one-dimensional bed support such as paper or TLC, or (ii) column chromatography with three-dimensional bed support. TLC is useful for separating mixtures of organic compounds and is often used to monitor the progress of organic reactions and to check the purity of products.

On the basis of physical state of mobile phase, chromatographic technique may be GC or LC. GC can be used to separate mixtures of volatile organic compounds. A GC consists of a flowing mobile phase, an injection port, a separation column containing the stationary phase, a detector, and a data recording system. LC is useful for separating mixtures of ions or molecules that are dissolved in a solvent. If the matrix support, or stationary phase, is polar (e.g., paper, silica, etc.), it is normal-phase chromatography; and if it is nonpolar (C-18), it is reversed-phase chromatography.

In short, chromatography is a method of separating the constituents of a solution, based on one or more of its chemical or physical properties. This could be charge, polarity, or a combination of these traits and pH balance. The solution is passed through a medium which will hinder the movement of some particles more than others. These principles are used to isolate and analyze enzymes, pigments, amino acids, constituents of DNA, and almost any other molecule you can imagine. A wide variety of chromatography techniques had developed to allow mixed substances to be separated [33].

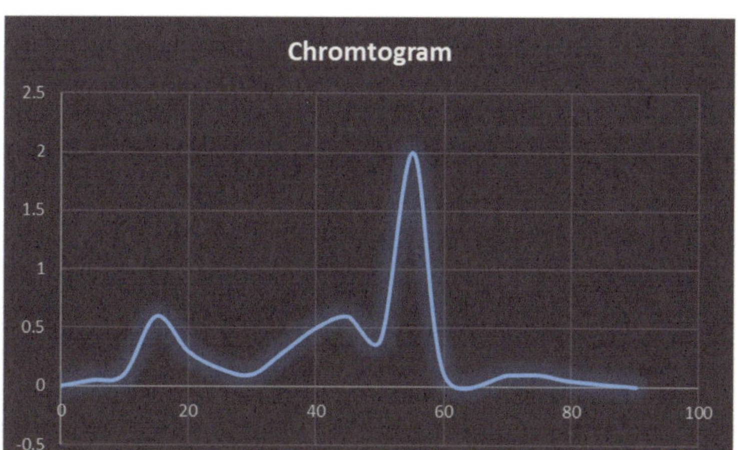

Figure 5.
Chromatogram response of a biological sample. The retention time is plotted on X-axis and signal on Y-axis obtained from detector corresponding to the response created by the analytes exiting the system.

7. Capillary electrophoresis (CE)

Jorgenson and Lukacs in 1981 invent capillary electrophoresis (CE) most often termed as capillary zone electrophoresis (CZE) [34]. It is a type of electrophoresis in which analytes are separated by applying an electric field across buffer solution-filled capillary tubes. The proposed instrumental technique was later implemented in a number of applications such as bioanalytical [35] and forensic drug analysis [36]. This technique is an alternative to the gel electrophoresis or LC (**Figure 6**).

CZE method correlated well with an automated kinetic fluorescent assay. An example is analysis of NAG by CE after incubation of urine samples using synthetic substrate, methylumbelliferyl-β-D-glucosaminide [37].

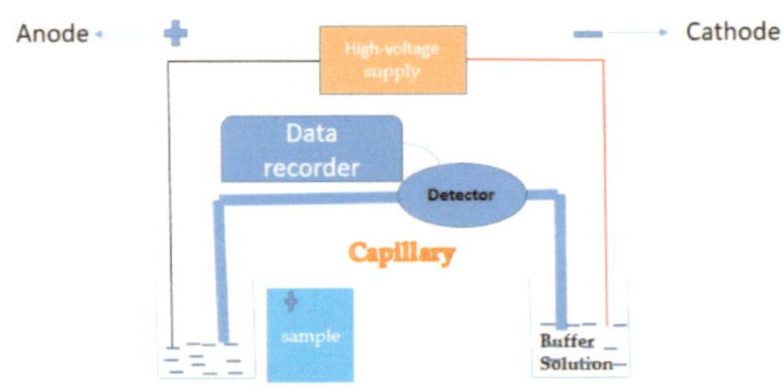

Figure 6.
Capillary electrophoresis separation method.

8. Microscopy

In the simplest microscopic methods, a specimen is illuminated by visible light and observed either against a bright background (bright-field microscopy) or a dark background (dark-field microscopy). The presence of cells that are not expected in the healthy person may be an indicator of disease. For example, a simple microscopic analysis of blood sample identifies the sickle cell anemia, and analysis of urine quantifies the presence of pus cells, which is an indicator of infection.

Light microscopy uses light as the illumination radiation. This is used to identify the microorganisms based on their morphology. An application of microscopy is to count the number of different cells per unit volume of blood or any other sample using a hemocytometer. Fluorescence microscopy has emerged as a very powerful tool for studying molecular processes owing largely to the advancement in optics and discovery of the green fluorescent protein and development of its analogs with different spectral properties [38]. Several advancements in the field of fluorescence microscopy have been achieved that includes the following techniques:

- Confocal laser scanning microscopy (CLSM)

CLSM is a type of fluorescence microscopy that allows imaging of the samples at different focal planes that light emitting from below or above the desired focal plane is eliminated. This results in very high lateral resolution and allows determining the spatial localization of the molecules [39].

- Total internal reflection fluorescence (TIRF) microscopy

TIFR is another type of fluorescence microscopy wherein the optics allows imaging of the molecules that are almost like microscopic slide. The resolution of light microscopes depends on the wavelength of the light used. The smaller the wavelength of the light used, the better the resolution obtained. Wavelength of the visible light imposes a resolution limit of ~0.2 μm on the light microscopes. Hemocytometer (Neubauer chamber) is a glass slide which has a counting chamber at the center. A glass cover is placed on the hemocytometer, and the sample is gently introduced into the chamber. The sample chamber has a grid which allows counting of cells in a defined region using a microscope [40].

9. Gene therapy protocol

Gene therapies are considerable improvements over the existing therapy because of the advantage in dosing schedule, patient compliance, toxicity, immunogenicity, and cost. Owing to this, gene therapy provides novel approaches for the treatment of inherited and acquired diseases. The development of a nonviral gene delivery vehicle capable of efficient, cell-specific delivery will be a valuable addition to the clinical armamentarium [41]. For example, the liver places a central role in the metabolism and production of serum proteins; it is an important target organ for gene therapy. Hepatic metabolic diseases as well as acquired diseases may also serve as targets for hepatic gene therapy [42]. Most recent gene therapy protocols describe delivery of foreign genes by means of injecting lentiviral particles [43].

10. Immunological and radioisotope techniques

Immunoassays are the quantitation of bioanalyte that depends on the reaction of an antigen (analyte) and an antibody. These methods are based on a competitive binding reaction between a fixed amount of labeled form of an analyte and a variable amount of unlabeled sample analyte for a limited amount of binding sites on a highly specific anti-analyte antibody. When immunoanalytical reagents (analyte or antibody) are mixed and incubated, the analyte is bound to the antibody forming an immune complex. This complex is separated from the unbound reagent fraction by physical or chemical separation technique. Analysis is achieved by measuring the label activity (e.g., radiation, fluorescence, or enzyme) in either of the bound or free fraction [44].

Immunoassay methods have been widely used in many important areas such as diagnosis of diseases, therapeutic drug monitoring, clinical pharmacokinetic and bioequivalence studies in drug discovery, and pharmaceutical industries. A few immunoassays based on different labels are as follows:

- Radioimmunoassay (RIA) methods have been used successfully for the determination of a limitless number of pharmaceutically important compounds in biological fluids. RIA is used to analyze thyroid hormone testing in patients after iodine-131 therapy.

- Enzyme immunoassay (EIA) is analogous to RIA except that the label is an enzyme rather than a radioisotope.

- Fluoroimmunoassay (FIA) is analogous to RIA except that the label is a fluorophore rather than a radioisotope.

- Chemiluminescence immunoassay (CLIA) involves a chemiluminescent substance as a label.

- Liposome immunoassay (LIA) is the assay involving a liposome-encapsulating marker.

- Cloned enzyme donor immunoassay (CEDIA) methodology is a novel approach which uses the DNA technology to produce homogenous enzyme immunoassays for drugs.

- Flow injection immunoassay (FIIA) methods were recently introduced to enhance the efficiency of immunochemical reaction, as well as to increase the performance of the analysis.

- Capillary electrophoresis immunoassay (CEIA) has been recently introduced as a sensitive analytical technique, particularly when combined with a sensitive detection method.

11. Hyphenated separation techniques and its application in clinical chemistry

The hyphenated techniques improved detection limits, sample identification capability, and miniaturization potential; hence, about 60% of the application of electrochemical detection (ED) has been found in the field of bioanalysis. The principle of ED used in biomedical analysis is a transfer of charge between substances in a column effluent and a working electrode; mainly, two types of ED either coulometric detection or amperometric detection are frequently used. The main advantages of using ED are the selectivity and sensitivity over UV-Vis detection. In HPLC, most of the application has been carried out by using UV-Vis detector, and ED is only used in small portion. The development of HPLC with ED facilitated highly sensitive and selective determination of homovanillic acid (HVA) and vanillylmandelic acid (VMA) in urine for the differential diagnosis of neuroblastoma pheochromocytoma and related tumors [45]. HPLC, coupled with UV-Vis using photodiode array as a detector, is widely used for determination of different drugs in serum. Other applications include determination of vitamins, antioxidants, and other components in biological samples.

CE coupled with MS provides an advantage of the sensitivity (parts per million range) and selectivity of these detection systems [46]. A detector that is becoming more frequently attached to CE is inductively coupled plasma mass spectrometry (ICP-MS). To date, CE-ICP-MS has been performed using a quadrupole detector within the MS allowing a small number of elements to be analyzed at any one time [47].

12. Lab on a chip (LoC)

Lab on a chip is defined as a microform of analytical devices that assimilate numerous laboratory operations such as PCR and DNA sequencing into a single chip on a very small scale. Miniaturized version of LoC provides cost-efficiency, use of low-volume reagents, high parallelization, high diagnostic speed, high sensitivity, and high expandability [48].

On the other hand, chronic disease (CD) healthcare is experiencing few limitations owing to lengthy and costly diagnosis procedures. Rapid, reliable, and low-cost diagnostic tools at point-of-care (PoC) instrumentation are therefore on high demand. LoC technology has a high potential to enable improved biomedical applications [49, 50]. In this regard, research toward developing new LoC-based PoC systems for CD diagnosis is fast growing into a nascent area such as chronic respiratory diseases (CRD), diabetes, and chronic kidney diseases (CKD) [51].

13. Conclusions

This chapter summarizes various non-separation and separation methods used for biomedical analysis. Although routine clinical methods indicate normal/abnormal levels of bioanalytes in urine/blood, still they often lack specificity due to severe complex biological sample interferences. In this regard, sample cleanup and highly sensitive techniques have proven to be helpful for early-stage disease diagnosis [52, 53] and detecting medical abnormalities [54]. To date, an array of hyphenated techniques plays an important role in the determination of bioanalytes, with improved selectivity and sensitivity.

Acknowledgements

I would like to sincerely thank Prof. R.S. Dubey and Prof. S.S. Pandey for their extensive support and guidance in the completion of this book chapter.

Conflict of interest

The author has declared no conflict of interest.

Thanks

I would also like to thank the staffs of the central library, Banaras Hindu University, Varanasi, India, for providing valuable books and journals.

Abbreviations

NAG	N-acetyl-beta-D-glucosaminidase
UV-Vis	ultraviolet-visible
IR	infrared
NIR	near infrared
BMV	bioanalytical method validation
LC	liquid chromatography
GC	gas chromatography
CE	capillary electrophoresis
MS	mass spectrometry
HPLC	high-performance liquid chromatography
SEM	scanning electron microscopy
CSF	cerebrospinal fluid
SPE	solid-phase extraction
SPME	solid-phase microextraction
MIP	molecularly imprinted polymer
CLSM	confocal laser scanning microscopy
TIRF	total internal reflection fluorescence
RIA	radioimmunoassay
FIA	fluoroimmunoassay
EIA	enzyme immunoassay
CLIA	chemiluminescence immunoassay
LIA	liposome immunoassay

CEDIA	cloned enzyme donor immunoassay
CEIA	capillary electrophoresis immunoassay
PCR	polymerase chain reaction
LoC	lab on a chip
PoC	point of care

Appendices

The *analyte* is the substance to be separated and to be found during chromatography from the mixture.

A *chromatograph* is an equipment that enables a sophisticated separation, for example, gas chromatographic or liquid chromatographic separation.

The output of the chromatograph is termed as *chromatogram*. In the case of an ideal separation, different peaks on the chromatogram correspond to different components of the separated mixture.

The *effluent* is the mobile phase leaving the column.

The *eluent* is the solvent that carries the analyte.

The *retention time* is the characteristic time it takes for a particular analyte to pass through the system (from the column inlet to the detector) under a set of conditions.

The *solute* is termed as the sample component in partition chromatography.

The *solvent* means any substance capable of solubilizing another substance, such as liquid mobile phase in LC.

The *detector* refers to an instrument used for quantitative and qualitative detection of analytes.

Author details

Khushaboo Pandey[1*] and Om Prakash Mishra[2]

1 Department of Biochemistry, Institute of Sciences, Banaras Hindu University, Varanasi, India

2 Department of Pediatrics, Institute of Medical Sciences, Banaras Hindu University, Varanasi, India

*Address all correspondence to: drkhushaboopandey@gmail.com

References

[1] Caroli S, Gyula Z. Analytical Techniques for Clinical Chemistry: Methods and Applications. Hoboken, N.J.: John Wiley & Sons, Inc.; 2012

[2] Beale DJ, Jones OAH, Karpe AV, Dayalan S, Oh DY, Kouremenos KA, Ahmed W, Palombo EA. A review of analytical techniques and their application in disease diagnosis in breathomics and salivaomics research. International Journal of Molecular Science. 2017;**18**(1):24

[3] Begum SS, Sushmaa BS, Vijayaraja S. Bioanalytical techniques—An overview. PharmaTutor. 2015;**3**(9):14-24

[4] Joseph L, Burtis CA, Bruns DE. Tietz Fundamentals of Clinical Chemistry and Molecular Diagnostics. 7th ed. 2008. Indian Journal of Clinical Biochemistry. 2015;**30**. DOI: 10.1007/s12291-014-0474-9

[5] Gad SC. Clinical Trials Handbook. Hoboken, N.J.: Wiley & Sons, Inc.; 2009

[6] Morgan JL, Lichtman JW. Digital tissue and what it may reveal about the brain. Biomed Central Biology. 2017;**15**:101-110

[7] Jia P. Plasmonic optical sensors: performance analysis and engineering towards biosensing [PhD theses]. The University of Western Ontario; Electronic Thesis and Dissertation Repository 2014;**2157**. Available from: https://ir.lib.uwo.ca/etd/2157

[8] Eghiaian F, Rico F, Colom A, Casuso I, Scheuring S. High-speed atomic force microscopy: Imaging and force spectroscopy. FEBS Letters. 2014;**588**(19):3631-3638

[9] Holland NT, Smith MT, Eskenazi B, Bastaki M. Biological sample collection and processing for molecular epidemiological studies. Mutation Research. 2003;**543**:217-234

[10] Pawliszyn J, Lord HL. Handbook of Sample Preparation. Hoboken N.J.: John Wiley & Sons Inc.; 2011. DOI: 10.1002/9780813823621.ch16

[11] Pungor E. A Practical Guide to Instrumental Analysis. Boca Raton: CRC Press LLC; 1995

[12] Sevick-Muraca E, Benaron D. OSA Trends in Optics and Photonics on Biomedical Optical Spectroscopy and Diagnostics. Washington, DC: Optical Society of America; 1916. pp. 271-272

[13] Xie J, Qian Z, Yang T, Li W, Hu G. Near-infrared spectroscopy technique to evaluate the effects of drugs in treating traumatic brain edema. Journal of Physics; Conference Series. 2011;**277**:012041

[14] Rao J, Butler DF, Taylor CR. Voorhees Van AS, Bissonnete R. Photodynamic Therapy for Dermatologists. https://emedicine.medscape.com/article/112157-overview

[15] Kartha VB, Santosh C. Biomedical Spectroscopy. India: Manipal University Press; 2014

[16] Severcan F, Haris PI. Vibrational Spectroscopy in Diagnosis and Screening. Amsterdam, Netherlands: IOS Press; 2012

[17] Rahman IU, Movasaghi Z, Rehman S. Vibrational Spectroscopy for Tissue Analysis (Series in Medical Physics and Biomedical Engineering). Boca Raton: CRC Press; 2013

[18] Driskell JD, Zhu Y, Kirkwood CD, Zhao Y, Dluhy RA, Tripp RA. Rapid and sensitive detection of rotavirus molecular signatures using surface enhanced Raman spectroscopy. PLoS One. 2010;**5**(4):e10222

[19] Magidson V, Khodjakov A, Digital Microscopy Methods in Cell Biology, 2013

[20] Jannetto PJ, Fitzgerald RL. Effective use of mass spectrometry in the clinical laboratory. Clinical Chemistry. 2016;**62**(1):1-7

[21] Xu W, Manicke NE, Cooks GR, Ouyang Z. Miniaturization of mass spectrometry analysis systems. JALA Charlottesv Va. 2010;15:433-439

[22] Keereweer S, Kerrebijn JD, Van Driel PB, Xie B, Kaijzel EL, Snocks TJ, et al. Optical image guided surgery—Where do we stand? Molecular Imaging and Biology. 2011;**13**(2):199-207

[23] Stradiotto NR, Yamanaka H, Zanoni Maria Valnice B. Electrochemical sensors: A powerful tool in analytical chemistry. Journal of Brazilian Chemical Society. 2003;**14**(2):159-173

[24] Ho WO, Krause S, McNeil CJ, Pritchard JA, Armstrong RD, Athey D, et al. Electrochemical sensor for measurement of urea and creatinine in serum based on AC impedance measurement of enzyme-catalyzed polymer transformation. Analytical Chemistry. 1999;**71**(10):1940-1946

[25] Stojanovic Z, Erdőssy J, Keltai K, Scheller FW, Gyurcsányi RE. Electrosynthesized molecularly imprinted polyscopoletin nanofilms for human serum albumin detection. Analytica Chimica Acta. 2017;**977**:1-9

[26] Serra PA. Biosensors. 1st ed. Vukovar, Croatia: InTech; 2010

[27] Gouvea. In: Serra PA, editor. Biosensors for Health Applications, Biosensors for Health, Environment and Biosecurity. London, U.K.: IntechOpen; 2011. pp. 71-86. DOI: 10.5772/16983

[28] Zhang L, Lv D, Su W, Liu Y, Chen Y, Xiang R. Detection of cancer biomarkers with nanotechnology. American Journal of Biochemistry and Biotechnology. 2013;**9**(1):71-89

[29] Atta NF, Galal A, Ali S. In: Serra PA, editor. Nanobiosensors for Health Care, Biosensors for Health, Environment and Biosecurity. London, U.K.: IntechOpen; 2011. DOI: 10.5772/17996 Available from: https://www.intechopen.com/books/biosensors-for-health-environment-and-biosecurity/nanobiosensors-for-health-care

[30] Ozkan-Ariksoysal D. Biosensors and their Applications in Healthcare. London U.K.: Future Science Group; 2013. Published Online. DOI: 10.4155/9781909453647

[31] Christian GD. Analytical Chemistry. 6th ed. Hoboken N.J.: John Wiley & Sons, Inc; 2016

[32] MLA style: Richard L.M. Synge – Facts. NobelPrize.org. <https://www.nobelprize.org/prizes/chemistry/1952/synge/facts/>

[33] Pandey K, Dubey RS, Prasad BB. A critical review on clinical application of separation techniques for selective recognition of uracil and 5-fluorouracil. Indian Journal of Clinical Biochemistry. 2016;**31**(1):3-12

[34] Jorgenson JW, Lukacs K, Arman D. Zone electrophoresis in open-tubular glass capillaries. Analytical Chemistry. 1981;**53**(8):1298-1302

[35] Zhong X, Zhang Z, Jiang S, Li L. Recent advances in coupling capillary electrophoresis-based separation techniques to ESI and MALDI-MS. Electrophoresis. 2013;**35**(9):1214-1225

[36] Anastos N, Barnett NW, Lewis SW. Capillary electrophoresis for forensic

drug analysis: A review. Talanta. 2005;**67**:269-279

[37] Friedberg M, Shihabi ZK. Analysis of urinary N-acetyl-β-glucosaminidase by capillary zone electrophoresis. Journal of Chromatography B, Biomedical Sciences and Applications. 1997;**695**(1):187-191

[38] Voura EB, Jaiswal JK, Mattoussi H, Simon SM. Tracking metastatic tumor cell extravasation with quantum dot nanocrystals and fluorescence emission-scanning microscopy. Nature Medicine. 2004;**10**(9):993-998

[39] Castellani C, Fedrigo M, Frigo AC, Barbera MD, Thiene G, Valente M, et al. Application of confocal laser scanning microscopy for the diagnosis of amyloidosis. Virchows Archiv. 2017;**470**(4):455-463

[40] Poulter NS, Pitkeathly WT, Smith PJ, Rappopor JZ. The physical basis of total internal reflection fluorescence (TIRF) microscopy and its cellular applications methods. Journal of Molecular Biology. 2015;**1251**:1-23

[41] Morgan JR. Gene Therapy Protocols: Reference Methods in Molecular Medicine. IInd ed. Vol. 69. New Jersey: Human Press Totowa; 2002

[42] Wu CH, Walton CM, Wu GY. Targeted gene transfer to liver using protein DNA complex. Methods in Molecular Medicine. 2002;**69**:15-23

[43] Buchschacher GL, Wong Stoul F. Bio artificial muscles. Development of lentiviral vectors for gene therapy for human diseases. Blood. 2000;**95**:2499-2504

[44] Darwish IA. Immunoassay methods and their applications in pharmaceutical analysis: Basic methodology and recent advances. International Journal of Biomedical Sciences. 2006;**2**(3):217-235

[45] Frattini P, Santagostino G, Schinelli S, Cucchi LM, Corona GL. Assay of urinary vanillylmandelic, homovanillic, and 5-hydroxyindole acetic acids by liquid chromatography with electrochemical detection. Journal of Pharmacological Methods. 1983;**10**(3):193-198

[46] Ramos-Payan M, Juan A, Ocana-Gonzalez RM, Llobera FA, Bello-López MA. Recent trends in capillary electrophoresis for complex samples analysis: A review. Electrophoresis. 2018;**39**:111-125

[47] Chetwynd AJ, Guggenheim EJ, Briffa SM, Thorn JA, Lynch I, Valsami-Jones E. Current application of capillary electrophoresis in nanomaterial characterisation and its potential to characterise the protein and small molecule corona. Nanomaterials. 2018;**8**(99):1-29

[48] Yılmaz B. Omics Technologies and Bio-Engineering Towards Improving Quality of Life. Elsevier B.V.; 2018. DOI: 10.1016/C2015-0-01634-3

[49] Herold KE, Rassoly A. Lab-on a Chip Technology: Fabrication and Microfluidics. London U.K.: Horizon Scientific Press; 2009

[50] Herold KE, Rassoly A. Lab-on a Chip Technology: Biomolecular Separation and Analysis. London U.K.: Horizon Scientific Press; 2009

[51] Wu J, Dong M, Rigatto LY, Lin F. Lab-on-chip technology for chronic disease diagnosis. npj Digital Medicine. 2018;**1**:7

[52] Prasad BB, Tiwari K, Singh M, Sharma PS, Patel AK, Srivastava S. Ultratrace analysis of dopamine using a combination of imprinted

polymer-brush coated solid-phase microextraction and imprinted polymer-sensor techniques. Chromatographia. 2009;**69**: 949-957

[53] Mishra OP, Rai AK, Srivastava P, Pandey K, Abhinay A, Prasad R, et al. Predictive ability of urinary biomarkers for outcome in children with acute kidney injury. Pediatric Nephrology. 2017;**32**(3):521-527

[54] Prasad BB, Tiwari K, Singh M, Sharma PS, Patel AK, Srivastava S. Zwitterionic molecularly imprinted polymer-based solid-phase microextraction coupled with molecularly imprinted polymer-sensor for ultratrace sensing of L-histidine. Journal of Separation Science. 2009;**32**:1096-1105

Chapter 4

Pre-Analytical Within-Laboratory Evacuated Blood-Collection Tube Quality Evaluation

Nataša Gros

Abstract

Pre-analytical steps contribute to an overall quality of the results of laboratory trials. The volume of blood drawn into the blood-collection tube and the anticoagulant amount introduced into the tube during its production should ensure the anticoagulant level in the recommended range; otherwise, the results can be altered. In evacuated blood-collection tubes, the internal under-pressure at the instant of the blood specimen collection affects the draw volume. During the shelf life, the internal under-pressure deteriorates. With no testing procedures in place, inappropriate anticoagulant levels can pass unnoticed. The chapter details testing procedures ensuring that the tubes are used only if, and only until, they are of the adequate quality. The reasoning behind the methodology is fully explained, and the case studies of the quality evaluations are discussed.

Keywords: evacuated blood-collection tubes, K_3EDTA, K_2EDTA, citrate, quality evaluation, draw-volume measurement, anticoagulant concentration

1. Introduction

Evacuated blood-collection tubes are evacuated containers intended for a venous blood specimen collection. They consist of a tube and a closure, which has to be tight to restrain low pressure—vacuum inside the tubes during their shelf life—but, on the other hand, it also has to be soft enough to let a sharp end of a blood-collection device to penetrate into a tube. The collection device has a disposable needle attached to the other side for phlebotomy.

Evacuated blood-collection tubes improved patients' and medical personnel's safety and mostly replaced classical tubes, which required a syringe and a needle for a specimen collection. In a continuation, wherever a tube is mentioned, the evacuated blood-collection tube is meant.

To prevent blood coagulation, tubes contain anticoagulants, either as dry substances attached to the internal walls or as solutions. Widely known and used are sodium citrate and salts of ethylenediaminetetraacetic acid (EDTA), usually present either as dipotassium or tripotassium salts, K_2EDTA or K_3EDTA.

Other substances, additives might be introduced as well to ensure the adequate properties or behavior of the tubes' internal walls or closures; nevertheless, they are expected not to interfere with a determination and affect analytical results. A noncompliant constituent detected in citrate tubes was magnesium which leached

IntechOpen

from a stopper and was consequently influencing the prothrombin time (PT) results [1]. A comprehensive study evaluating different tubes comprising also recently introduced low-magnesium version confirmed that the PT and INR differences between the tubes are correlated with the magnesium concentration differences [2].

Manufacturers are obliged to specify on the label of a tube: a type of anticoagulant, a nominal draw volume, a lot number, and expiration date; within this text, we use a term expiry date as well.

Anticoagulant concentration in a blood sample after specimen collection should be within an appropriate range; otherwise, analytical results might be altered. To reach this objective, an accurate amount of anticoagulant should be introduced into a tube during production, and a draw at the moment of a specimen collection should be adequate to ensure a volume of blood entering a tube is within an acceptable range. A label on a tube provides guidance for inspection if the volume is within the suggested limits; however, this is only true if the label is precisely and accurately positioned.

The latest version of the GP39-A6 standard of the CLSI standardization body (Clinical and Laboratory Standards Institute) [3] requires of the tubes' manufacturers to ensure that until expiration date, the anticoagulant concentration remains within the 5% range of the value stated on a label. A draw volume is considered acceptable if it does not differ from the stated nominal volume for more than $\pm 10\%$.

The standard GP34-A recognizes the importance of appropriate blood-to-EDTA ratio for obtaining optimal examination results but avoids stating the exact limits. The EDTA can, if in a concentration which is too high hypertonically shrink red cells, affect red cell size and cause morphological changes. On the other hand, it can too extensively chelate calcium and other cations such as magnesium and zinc and affect the activity of alkaline phosphatase enzyme label used in chemiluminescent assays or reduce the efficiency of the recognition of proteins by antibodies due to the proteins' conformation changes [4].

The predecessor of CLSI, the National Committee for Clinical Laboratory Standards (NCLLS), was in the H1-A5 standard more explicit in terms of some anticoagulants' concentrations [5]. It explains that only a little bit less than a half (1.15 mmol/L) out of the total calcium concentration (2.5 mmol/L) corresponds to unbound calcium that needs to be chelated stoichiometrically with EDTA to prevent coagulation. For that reason, it suggests that EDTA concentration in blood should be between 3.7 and 5.4 mmol/L, since excessive concentration causes morphological changes in the blood.

Not consistent with this requirement was DIN ISO 6710: 1996-12 standard requiring the EDTA concentration within the 4.11–6.843 mmol/L range [6].

A potential user can during time come across the tubes which were produced by not having the same set of requirements on the mind. As we already previously demonstrated, the tubes if evaluated as such not yet in contact with a blood sample are not all the same, and change in their own characteristics during their shelf life and the testing procedure which we suggested are easy to perform [7].

A concise review reflects on the behavior of EDTA as an anticoagulant in hematology and furthermore discusses its usage in proteomics, general clinical chemistry, and its applicability for measuring cytokines, protein, peptides, and cardiac markers [8]. Elsewhere, influences of a form of EDTA and its concentration on the results of hematological tests were profoundly discussed in relation to spurious counts and results regarding platelets [9], white blood cells, red blood cells, hemoglobin, red cell indices, and reticulocytes [10]; under-filled or over-filled evacuated tubes changing the anticoagulant level in a sample are exposed as an influential pre-analytical source of errors.

For citrate tubes the DIN ISO 6710: 1996-12 standard recommends trisodium citrate solutions with concentrations between 100 and 136 mmol/L; however, the H1-A5 standard specifies the concentrations 105, 109, and 129 mmol/L.

Due to all these differences, the GP39-A6 standard omitted all the anticoagulant concentrations' details, leaving it entirely to a producer to bear the responsibility for securing appropriate concentration, fulfilling all the requirements, and demonstrating that they are actually met, or in other words verifying that the tubes are actually fit for purpose.

The GP34-A standard provides guidance for validation and verification of tubes for venous and capillary blood specimen collection [4]. Both a manufacturer and a clinical laboratory are required to perform a comparability study on blood samples for two or more sets of tubes comprising a set which was already evaluated and approved previously. A manufacturer performs such a test after a new product was developed or where any correction actions are necessary for the production process. The laboratory needs to do it when switching from one product to another or when changing a vendor.

A within-tube precision study requires a minimum of 20 subjects, and each sample needs to be analyzed in replicates; an appropriate number of samples, evenly distributed through the analytical measurement, are essential for trueness evaluation [4]. Several studies with sometimes dissimilar outcomes can be found in the literature.

Two blood-collection devices either with an aligned [Becton Dickinson (BD)] or at an angle needle holder (Greiner Labortechnik GmbH) were evaluated either enabling a direct linear (BD) or interrupted nonlinear blood flow. A mechanical strain on blood cells was recognized as a factor potentially causing the efflux of intracellular constituents into the serum in an interrupted nonlinear flow. The magnesium, plasma hemoglobin, and prothrombin time within-subject variations were confirmed in 55 healthy individuals using a Student paired t-test. A difference in a tube material either glass or polymer was also recognized as a likely contributing factor [11].

Nevertheless, contrasting outcomes were obtained for prothrombin time determinations in the glass and PVC tubes with two distinct citrate concentrations where neither material caused the significantly different results [12]. Yet another study, establishing a protocol for comparing the citrate evacuation blood-collection tubes with glass tubes employing eight measuring systems, confirmed a statistically significant but clinically not relevant difference in prothrombin time results, which were more pronounced with the tubes of the lowest 2.7 mL draw volume [13].

Differences in some parameters were confirmed if BD plastic citrate tubes were used instead of glass tubes, but they were considered unlikely to be clinically significant [14], though a comprehensiveness of a study was challenged arguing that only healthy volunteers were involved and by these means it was not yet proven that glass tubes are interchangeable with the plastic tubes [15]. But a study performed on Greiner glass citrate and plastic tubes confirmed that the tubes are substitutable as far as either untreated or patients on a traditional oral anticoagulant therapy are concerned and that this applies for the whole shelf life of the tubes [16].

Nevertheless, the plastic tubes of different brands evaluated on patients and healthy volunteers were confirmed to be statistically but not clinically significantly different [17]. For patients on oral anticoagulant therapy with vitamin K antagonists, ANOVA test confirmed statistically significant differences in prothrombin time for the tubes of four different types [18]. The study supports the claim that validation is always necessary when there is a change in a tube type.

A research performed on a group of individuals evaluating the effect of underfilled EDTA tubes on hematological parameters by employing a particular type of analyzer [19] leads to contrasting outcomes not necessarily aligned with other studies [20] and general principles and recommendations.

Validations and verifications as required by the standard GP34-A are complex to perform, time demanding, and require resourceful personnel [4].

The standard exposes a blood collection as a pre-analytical (preexamination) source causing varying degrees of errors. It brings to light a lack of a mechanism that would enable systematic evaluations of the influences of pre-analytical (preexamination) variables on laboratory test (examination) results [4].

The characteristics of the tubes entering the pre-analytical phase are such variables, and this is where this chapter tends to contribute.

Differences between tubes of different brands examined 5 years apart in time are going to be enlightened, and the testing procedures which are fast, cheap, and easy to implement into laboratory practice are explained in full details. Robustness of personal profiles of athletes and validation studies performed on blood samples can profit from knowing the attributes of the tubes that were actually used or evaluated.

2. Quality of evacuated blood-collection tubes for hematological tests reevaluated in 5 years' time

In this section, we compare the results of a quality evaluation of the blood-collection tubes, on which we previously reported [7], with the results obtained in 5 years' time. The same producers were considered, namely, Becton Dickinson, Greiner Bio-One, and Laboratorijska tehnika Burnik d. o. o.

Figures 1–4 are dedicated to the tubes of four different brands. The abscise axis stands for a draw volume. Ordinate on the right indicates an anticoagulant concentration expected for a blood sample after a specimen collection. We are going to explain the meaning of the left ordinate axis, which relates to a testing procedure, later. A frame in green indicates the limits set by the H1-A5 standard [5]. The horizontal lines confine the range of the acceptable anticoagulant concentration; the left vertical line represents a limit under which a draw volume is expected not to fall to prevent the anticoagulant concentration to rising too high.

We kept the assigned marks A, B, and C from the previous study for the K_3EDTA tubes, and D for the K_2EDTA tubes, not disclosing the producers' identity. We additionally included the tubes F, not previously evaluated. The more recent

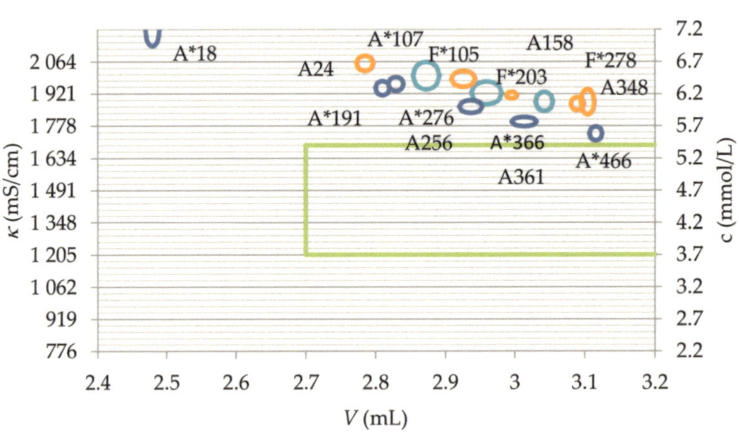

Figure 1.
Anticoagulant concentrations and draw volumes of the A brand tubes obtained 5 years apart in time (orange/blue) and of the tubes F not previously included (the numbers in the labels indicate time until the expiration).

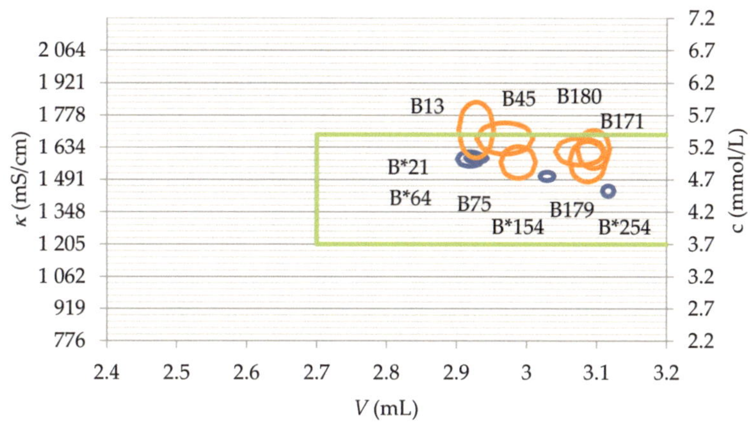

Figure 2.
Anticoagulant concentrations and draw volumes of the B brand tubes obtained 5 years apart in time (orange/blue); the numbers in the labels indicate the time until the expiration.

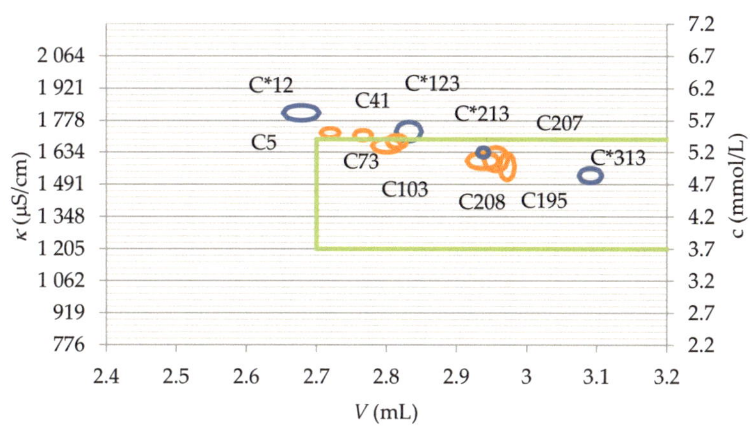

Figure 3.
Anticoagulant concentrations and draw volumes of the C brand tubes obtained 5 years apart in time (orange/blue); the numbers in the labels indicate the time until the expiration.

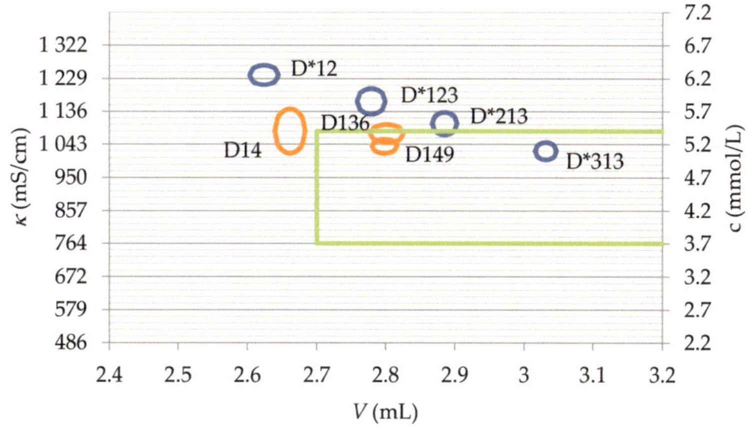

Figure 4.
Anticoagulant concentrations and draw volumes of the D brand tubes obtained 5 years apart in time (orange/blue); the numbers in the labels indicate the time until the expiration.

results are marked with an asterisk; a number indicates the number of days until the expiration.

Ellipses in **Figures 1–4** provide an insight into a spread of results obtained for the tubes within a series of measurements. Ellipses in blue correspond to the more recent results; those in orange originate from a previous study. A length of a horizontal axis of an ellipse equals a standard deviation of the draw-volume measurements; a vertical axis indicates a standard deviation of the anticoagulant concentration as expected for blood samples. A crossing of the two axes of an ellipse is defined by the mean values of the two parameters. The smaller the ellipse, the higher the quality of the produced tubes in terms of their precision or, in other words, a repeatability of a product.

2.1 K_3EDTA tubes

From this point of view, the A brand tubes were and remain of a high quality. In spite of having a very long expiration date, unless very close to the expiration (A*18), the draw volumes generally remain adequate; however, the anticoagulant concentration is too high and not in the accordance with the higher limit set by the H1-A5 standard [5]. All the ellipses of the A brand tubes are above the green rectangle in **Figure 1**. The same is true for the tubes F, which are of the same manufacturer. This example demonstrates that an adequate draw volume does not yet ensure the adequate anticoagulant concentration as far as the H1-A5 is concerned. The tubes were obviously produced following an older DIN ISO 6710: 1996-12 standard, for which the anticoagulant concentration 6.843 mmol/L was still acceptable. With this limit in mind, only the A*18A anticoagulant concentration would be excessive. The characteristics of these tubes clearly contrast those of the tubes B and C, as the figures in the continuation demonstrate.

In the B brand tubes (**Figure 2**), an improvement in the overall quality of the results is evident in 5 years' time. Previously the ellipses were much larger, and they exceeded the upper anticoagulant concentration limit even with the draw volumes very close to the nominal 3 mL mark. The ellipses in blue are small and remain within the green rectangle during the whole shelf life; nevertheless it has to be mentioned the shelf life was at the time of purchase much shorter than in the A brand tubes.

The C brand tubes exhibit a good repeatability of the product; however, the anticoagulant concentration starts exceeding slightly the higher-anticoagulant concentration limit already with the draw volumes approaching 2.8 mL, proving that the draw volume within the acceptable range does not yet guarantee the correct anticoagulant concentration.

2.2 K_2EDTA tubes

The product repeatability of the D brand tubes is good, but the anticoagulant concentration starts exceeding the upper limit already approximately 200 days before the expiration date and with the draw volumes falling below 2.9 mL what is far above the acceptable lower limit.

This study confirmed that a control of a draw volume is not the main quality issue and does not ensure that the anticoagulant concentration is adequate. Only in four cases, the draw volumes below the 2.7 mL limit were observed, and this only happened when the tubes were tested closer than 20 days before their expiration date. On the other hand, in more than 20 cases, the anticoagulant concentration limit as set by the H1-A5 standard was exceeded at the draw volumes in the recommended range.

It also needs to be mentioned that if the draw-volume inspection relays on the label mark a judgment can be false. During our first study, we found out that only one brand of the tubes had a label positioned precisely; in all others the indicators on the tube were misleading.

Easy-to-perform testing procedures as we used here which do not require blood samples can as a precautious measure ensure that the tubes are used only if they are of the adequate quality and their quality does not fluctuate too much during the time. It can alert a laboratory when it would be advisable to perform a much more complex and time-demanding verification study on blood samples. Archived data on the tubes' characteristics and quality during a longer period of time can provide a piece of evidence for other studies and rule the tubes out as a potential cause for variations.

2.3 Tubes' drawing capability reduces during the time

In this section, we explain the changes in a behavior of the tubes during their shelf life.

The tubes' drawing capability depends on a difference between the external (p_{st}) and internal ($p_{int\_20°C}$) pressure. The lower the internal pressure, and the higher the difference to the external pressure, the higher a drawing capability. A tube's internal volume (V_{tube}) also contributes to higher capacity to draw a liquid.

The tubes of different brands differ in their drawing capability and, in a way, how it reduces during the time, as **Figure 5** demonstrates. No container is entirely tight and leaks to some extent. The conditions to which the tubes were exposed or under which they were stored contribute. The tubes of the same lot would behave differently in different circumstances.

Even though tubes are of a high quality, are purchased at the same time, and are of the same lot, they are not all the same if used during their shelf life.

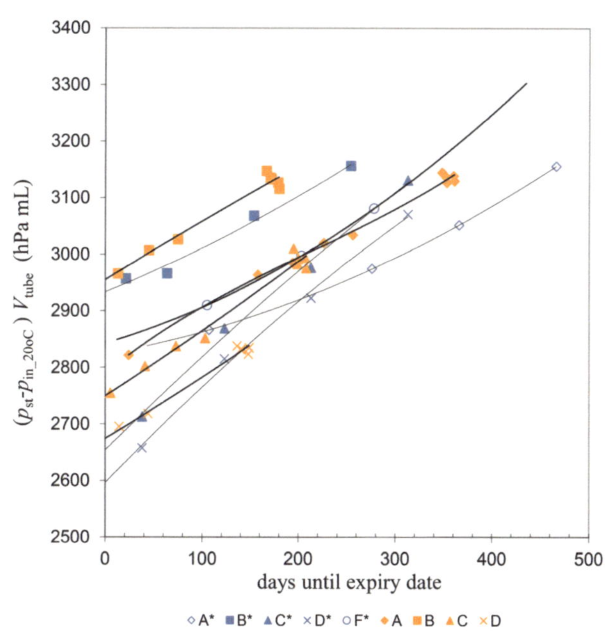

Figure 5.
Drawing capability of the tubes reduces continuously during their shelf life.

A reduction in a drawing capacity results in a diminished draw volume and enhanced anticoagulant concentration.

The tubes' validations and verifications with blood samples are essential; however complex and with no insight into characteristics of the tubes as such, it is not clear what was actually tested and how representative it is. The outcomes might vary and can be influenced by a choice of a group of individuals, its representativeness, a normality of a distribution of the investigated parameter, and the sources of uncertainty originating from the whole procedure, comprising pre-analytical, analytical, and post-analytical phases. All these factors can influence the conclusions of a paired t-test or ANOVA, a difference between two brands, or lots of the tubes might turn out insignificant. Insight into tubes' characteristics can give the verification and validation studies some basic orientation on what was the status of the tubes that were investigated. If the evaluation is repeated later, one can know how comparable are the examined tubes entering the process.

If a difference between the results obtained for the tubes of different brands is confirmed to be statistically significant by the tests performed on blood samples, they are frequently considered not clinically important. However, variations, which are not important on the level of a group of individuals, might reflect differently when a single person is concerned. Personal variations in blood parameters are narrower than variations on the level of a population. A personal medicine and athlete's biological passport require higher sensitivity and attention paid to all sources of uncertainty.

It is not possible to perform validation study with blood samples for each individual, but it is easily possible to perform a quality control of the tubes which are used for personal profiles.

The athlete's biological passport (ABP) requires accurate and reliable results of hematological tests, which are stable enough for evaluations of the probabilities of abnormalities in a personal profile. Hence, some parameters, e.g., hemoglobin and erythrocytes, were identified as highly stable; the others such as reticulocytes, mean red blood cell volume, and hematocrit did not turn out as such. Sample storage conditions and treatment and the choice of an analyzer are considered the contributing factors [21].

It was proven that during a training season hemoglobin and hematocrit reduce in their value, and reticulocytes do as well but independently. The pattern is general but the size of a change is sport's discipline dependent. It was recognized that reliable reference ranges in sportsmen could not be defined without the best laboratory practices [22].

Not univocal and entirely clear outcomes of different studies on the stability of the blood variables raise concerns and request for more clearly defined characteristics, procedures, threshold limits, personal reference ranges, and criteria for recognizing abnormalities to prevent false convictions in athletes [23].

In order to raise awareness to which extent different pre-analytical phases could affect the outcomes of hematological and biochemical tests on which sports medicine depends in following athletes, different pre-analytical aspects and the choice of anticoagulant, instabilities of some molecules were addressed to prevent misinterpretation of data and improve the usefulness of results [24]. Specimen homogenization as a pre-analytical phase received special attention [25].

The variations in tubes' characteristics did not receive attention in relation to ABP in spite of the fact that easy-to-perform testing procedures are available [7]. They are thoroughly explained in the continuation.

3. A methodology for K_3EDTA or K_2EDTA evacuated blood-collection tubes' quality evaluation

A methodology for a quality evaluation of evacuated blood-collection tubes for hematological tests consists of two successive measurements, a measurement of a draw volume and electrolytic conductivity, from which one can predict the anticoagulant concentration in a blood sample. No patient- or person-related samples are required. A medium for the tests is purified water.

Only low-cost equipment, a Bang burette and a field conductivity meter, is used. One also needs to know an ambient temperature and a non-reduced pressure for a period during which measurements were performed. The latter can be obtained from a local meteorological institution on request, and temperature is easy to measure. We explain the testing procedures in full details in the following section.

We previously published the nomograms for K_2EDTA and K_3EDTA tubes with nominal 3 mL draw volume, which enable a prediction on how is an anticoagulant concentration going to rise with a diminishing draw volume that happens during the time because of the aging of the tubes [7].

In **Figure 6** we present a nomogram for the K_3EDTA tubes with a 2 mL nominal draw volume relating the electrolytic conductivity, κ, with a predicted anticoagulant concentration, c, and a draw volume, V. The points are the results of the quality evaluation of the tubes of two different brands, E and G, at the particular moment in time. If the tubes in a lot are of a homogenous quality and they are going to be tested later during their shelf life, the anticoagulant concentration as expected for blood samples is going to rise as the curves indicate. In such a case, a draw-volume measurement can already give an insight into the quality of the examined tubes.

If we take the G tubes as an example, for the great majority of the tested tubes, with a draw volume close to 2.1 mL, the anticoagulant concentration anticipated initially was 4.45 mmol/L. If we assume that later in time we find out that the draw volume has fallen to 1.9 mL, the anticoagulant concentration for blood samples is going to be close to 4.95 mmol/L. A green rectangle indicates the limits set by H1-A5 standard [5], demonstrating that the tubes would still have been of adequate quality.

Skills needed to perform the tests are not difficult to master, and a laboratory or medical staff can easily develop them. The quality of the tubes does not deteriorate very rapidly. It is important to test tubes when they are put into use, and later only

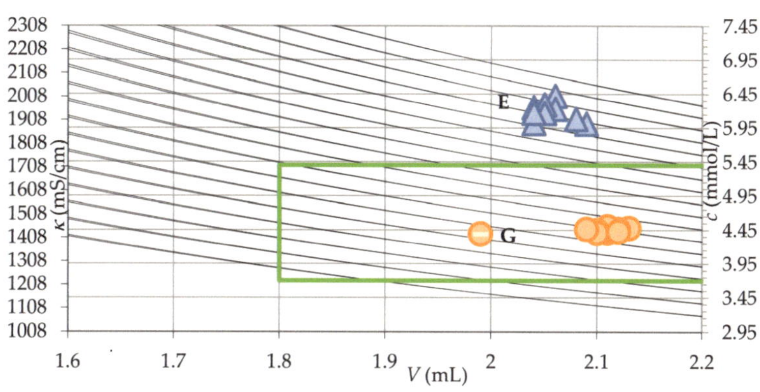

Figure 6.
A nomogram for a prediction on how is the anticoagulant concentration in 2 mL K_3EDTA tubes of two different brands (E, G) going to change with a declining draw volume; green rectangle outlines the characteristics considered acceptable by H1-A5 standard [5].

occasionally, but at regular intervals. Since the tests are not time-consuming and not performed in high numbers, this additional workload does not represent an important additional burden for personnel; however, the benefits for an institution are obvious and important.

An institution implementing a quality evaluation scheme always has an adequate insight into the characteristics and quality of the tubes it is using. It can ensure that the tubes are used only if and only until they are of adequate quality or it can use the data in medical and clinical studies to test possible correlations or covariations.

4. Testing procedures

Testing procedures consist of a draw volume and an electrolytic conductivity measurement (**Figure** 7 left/right), both later corrected to correspond to a temperature 20°C and an external pressure 101 kPa as required by the standards [5, 6].

4.1 Draw volume

In the schematic (**Figure** 7), far left, an evacuated tube, characterized by an internal pressure (p_{int}) and an internal tube volume (V_{tube}), defining a conserved energy of withdrawal for a blood specimen collection is depicted.

In the middle, a draw-volume measurement is schematically represented. A starting point is a Bang burette filled with purified water to the 0 mL mark. A tip of the burette is attached to a flexible tubing, which is at the other end connected to a blood-collection device. This part too is entirely filled with purified water. When we attach an evacuated tube to the venipuncture device, the tube starts filling with water, and consequently the water level in the Bang burette starts falling. The rising water level in the tube acts as a moving piston, reducing the void volume and causing the internal pressure to rise until it equals the external pressure (p_{ext}). At this moment, the withdrawing ends, and we can read the draw volume from a burette.

Hence, the external pressure and temperature (T) affect a draw-volume test. The external pressure depends on the altitude and current weather conditions and can be obtained from the local meteorological institution. The ambient temperature is also important and has to be taken into account. It affects the internal pressure in

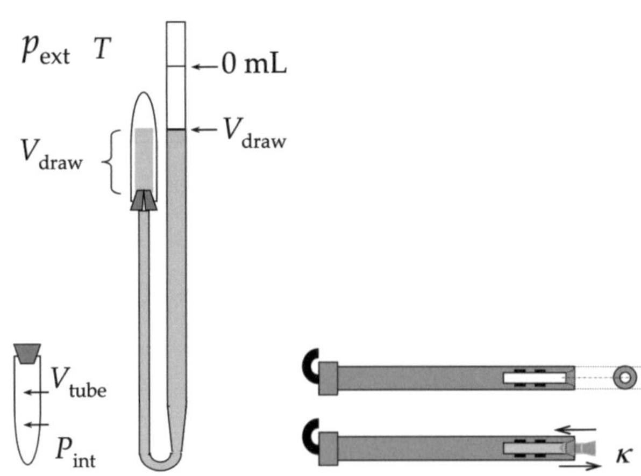

Figure 7.
A setup for a draw-volume measurement (left) and a conductivity cell for the anticoagulant concentration estimation (right).

a tube at the instant of the draw-volume measurement. At a higher temperature, the air in the tube expands, the internal pressure rises, or, in other words, the internal under-pressure deteriorates, and a tube's withdrawing ability falls.

For these two reasons, a draw volume (V_{draw}) should be corrected to obtain an estimation of a draw volume, as it would have been, if measured at 1013 hPa and 20°C (V_{draw_st}) (1); the symbol K and hPa stand for the units Kelvin and hectopascal, respectively:

$$V_{draw\_st} = V_{tube} - \frac{p_{ext} \times 293.16 \times \text{K}}{T \times 1013 \times \text{hPa}}(V_{tube} - V_{draw}) \qquad (1)$$

The standards [5, 6] admit that if the draw volume under these conditions does not differ from the nominal volume for more than 10%, the tubes are expected to be of the adequate quality for a blood specimen collection. For the 3 mL tubes, this means that the draw volumes between 2.7 and 3.3 mL are acceptable.

Even though the standards [3, 4] expose as the main quality issue in a nonconformity of a draw volume with the requirements, our results in Section 2 prove that a draw volume within the acceptable range does not necessarily ensure the correct anticoagulant level. An additional insight into this aspect of quality is necessary. For the K_2EDTA and K_3EDTA tubes, respectively, a conductance measurement can provide such an insight.

4.2 Ionic conductivity

K_2EDTA and K_3EDTA are salts, and salt in water dissociates in ions. Solutions containing ions conduct electric current, to which extent depends on the ions' characteristics and their concentration. In other words, conductance (G) of a solution reflects an overall ionic composition, but, if a solution contains only a single salt, as it is a case for K_2EDTA and K_3EDTA tubes, it can provide an insight into a salt concentration.

A conductance depends on the geometry of a conductivity cell. A cell we used is depicted in **Figure** 7 (far right). It was an immersive four-electrode cell, not directly applicable to our needs. We closed the bottom of the cave with a parafilm and the hole at the right with a stopper made of a pipette tip to transform it into a conductivity cell applicable for conductance measurements in solutions of a small volume.

The conductance measurement is affected by a size of electrodes and a distance between them; a cell geometry is reflected in a cell constant K. If the results of measurements are expressed as ionic conductivity (κ), they are universally useful and comparable between different measuring systems (2):

$$\kappa = G \times K \qquad (2)$$

Another concern is a temperature during a measurement; it also affects the conductance value. At higher temperatures, a conductance is higher. This is taken into account by reporting an ionic conductivity at a selected reference temperature, e.g., 20°C. Measurements obtained at an ambient temperature are transformed into the values as would have been at the reference temperature by taking a temperature compensation into account. A linear compensation is frequently used; a compensation factor is usually approximately 2%/°C. In our case, the values confirmed experimentally were 1.99 or 2.01% for K_3EDTA or K_2EDTA, respectively.

Conductivity measurements are in fact easy to perform. A conductivity cell has a temperature sensor incorporated. We select the reference temperature, define a temperature compensation factor and a cell constant, and perform measurements.

4.3 Anticoagulant concentration

For a solution of a single salt, electrolytic conductivity values are easy to transform into a salt amount concentration, *c*. If one prepares a set of the solutions with the known salt concentrations and determines their electrolytic conductivity at the reference temperature, one can relate the two parameters by depicting a graph, with the first parameter on the abscise axis, and the second on the ordinate, defining a trend line (3):

$$\kappa = a + b \times c \tag{3}$$

The symbols, *a* and *b*, are the parameter of the linear equation; *a* stands for the intercept and *b* for the slope.

After these two parameters are known, one can measure the electrolytic conductivity of the anticoagulant solution obtained after a draw-volume test to obtain κ_{V_draw} and use the equation in a rearranged form to calculate a concentration of the anticoagulant, c_{V_draw} (4):

$$c_{V\_draw} = \frac{\kappa_{V\_draw} - a}{b} \tag{4}$$

This concentration is valid at the draw volume, but one wants to know the anticoagulant concentration as expected in blood after a specimen collection, $c_{V\_draw\_st}$. As already explained, the draw volume had to be corrected to obtain an estimation of a blood sample volume; a concentration is volume dependent, and therefore it has to be corrected too, to correctly represent the anticoagulant concentration as expected after a blood specimen collection (5):

$$c_{V\_draw\_st} = c_{V\_draw} \frac{V_{draw}}{V_{draw\_st}} \tag{5}$$

Draw volume and anticoagulant concentration determination can be implemented into laboratory quality control routines. If followed during the time in a form of control charts, they can ensure that the tubes are used only if and only until they are of adequate quality.

5. Citrate tubes a distinct case

The same testing procedure for a draw-volume measurement is applicable also to citrate tubes. However, in terms of a determination of the anticoagulant concentration, this is a distinct case. The reason is that citrate in the tubes can be either buffered or unbuffered. As a result, a solution after a draw-volume test can have quite a different pH and contains different citrate equilibrium forms in different proportions. For this reason measurement of electrolytic conductivity cannot be used here, and an adequate low-cost and easy-to-perform testing procedure yet has to be developed.

We evaluated the citrate tubes of two different brands, A and B. Both were (1:9) type tubes but differed in nominal draw volumes, which were 4.5 and 3.6 mL for tubes A and B, respectively. Measured draw volumes that we obtained were 4.55 and 3.47 mL; it has to be pointed out that these are uncorrected values.

Getting insight into the composition of the anticoagulant solution after a draw-volume test is possible, but the approach is complex and time-consuming. We used

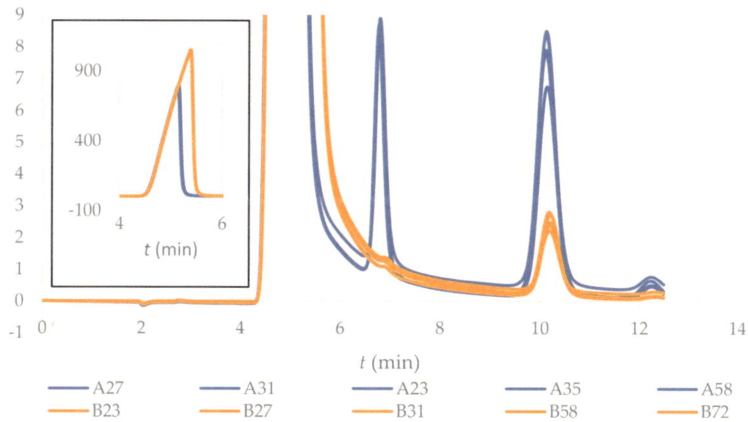

Figure 8.
Chromatograms for determination of sodium, potassium, magnesium, and calcium in the tubes of two different brands, A and B, after a draw-volume test.

two ion chromatographic techniques, ion-exclusion chromatography for determination of citrate and ion-exchange chromatography for determination of cations.

Figure 8 depicts chromatograms obtained for determinations of cations after a draw-volume test. Four chromatographic peaks emerge. The first peak appearing between 4 and 6 minutes pertains to the major cation, sodium, originating from the anticoagulant, trisodium citrate. This peak is presented twice with two different y-axis scales, in the main chromatogram with its top is out of scale and in the inserted frame where it is fully visible.

The peak for the A brand tubes is smaller, indicating that in this case citrate is buffered, while in addition to sodium ions there are hydronium ions (H^+, or H_3O^+) to neutralize a negative charge of citrate equilibrium forms. In other words, in the case of the A brand tubes, the anticoagulant solution was prepared not only from a trisodium citrate but also with an addition of citric acid. Consequently, the pH of a solution is lower than with the B brand tubes; the pH results we obtained were 6.1 and 7.9, correspondingly.

As **Figure 8** demonstrates, we also confirmed the presence of potassium, a peak between 6 and 8 minutes; magnesium, a peak at around 10 minutes; and calcium, a peak at around 12 minutes.

Presence of magnesium came with no surprise; authors using a different analytical method previously reported on it and explained that it leaks from a closure and originates from an additive [1]. Though we confirmed a difference between A and B brand tubes, the latter contained magnesium in much lower concentration.

The concentration of calcium is low in both cases and might be considered an impurity originating from other chemicals. But what makes a real difference between the tubes of the two brands and was not expected is potassium. While it is nearly inexistent in tube B, it is obviously present in tube A. We used infrared spectroscopy to explain where it originates from; it appears that the source might be a tripotassium salt of EDTA; if this is the case, its concentration is approximately 200 times lower than the citrate concentration.

6. Conclusions

This research proved differences in characteristics of tubes of different brands and different lots, and their attributes are also changing during a shelf life. We suggested fast, easy-to-perform testing procedures, which already by using purified water and low-cost equipment only give an insight into draw volume and anticoagulant concentration as can be expected for a blood specimen collection.

No doubt, a laboratory has to prove on blood samples that a particular type of tubes it is using or intends to use is fit for purpose. But already with a minimum required number of samples of 20 individuals, a study becomes complex, professionally demanding, and time-consuming. However, not knowing the characteristics of the tubes entering the investigation lacks generality.

Testing procedures as suggested are not to replace but to support such investigations and to make them more economical. A quality control of the tubes as such is easy to introduce into a laboratory practice and does not importantly contribute to a workload. The tubes do not change very rapidly over the time and do not need to be tested very frequently if of the same lot. However, insight into their characteristics provides guidance when a study on blood samples needs to step in to cover important distinctive conditions a laboratory is likely to face during its everyday routine.

A draw-volume test we described is generally applicable. Procedures for determining K_2EDTA and K_3EDTA concentrations in the tubes before a specimen collection are well established; nevertheless comparably easy-to-perform testing procedures for citrate tubes yet need to be developed.

Acknowledgements

A financial support of the ARRS research program P-153a is acknowledged.

Conflict of interest

No conflicts of interest are declared by the authors.

Author details

Nataša Gros
Faculty of Chemistry and Chemical Technology, University of Ljubljana, Ljubljana, Slovenia

*Address all correspondence to: natasa.gros@fkkt.uni-lj.si

References

[1] van den Besselaar A, van Dam W, Sturk A, Bertina RM. Prothrombin time ratio is reduced by magnesium contamination in evacuated blood collection tubes. Thrombosis and Haemostasis. 2001;**85**(4):647-650

[2] van den Besselaar A, van Vlodrop IJH, Berendes PB, Cobbaert CM. A comparative study of conventional versus new, magnesium-poor Vacutainer (R) sodium citrate blood collection tubes for determination of prothrombin time and INR. Thrombosis Research. 2014;**134**(1):187-191. DOI: 10.1016/j.thromres.2014.04.016

[3] CLSI. Tubes and Additives for Venous and Capillary Blood Specimen Collection. Approved Standard Sixth Edition. CLSI document GP39-A6. Wayne, PA: Clinical and Laboratory Standards Institute; 2010

[4] CLSI. Validation and Verification of Tubes for Venous and Capillary Blood Specimen Collection. Approved Guideline. CLSI document GP34-A. Wayne, PA: Clinical and Laboratory Standards Institute; 2010

[5] NCCLS. Tubes and Additives for Venous Blood Specimen Collection. Approved Standard Fifth Edition. NCCLS document H1-A5. Wayne, Pennsylvania: NCCLS; 2003

[6] DIN ISO 6710: 1996-12 Single-Use Containers for Venous Blood Specimen Collection. 1996

[7] Gros N. Evacuated blood-collection tubes for haematological tests—A quality evaluation prior to their intended use for specimen collection. Clinical Chemistry and Laboratory Medicine. 2013;**51**(5):1043-1051. DOI: 10.1515/cclm-2012-0507

[8] Banfi G, Salvagno GL, Lippi G. The role of ethylenedimine tetraacetic acid (EDTA) as in vitro anticoagulant for diagnostic purposes. Clinical Chemistry and Laboratory Medicine. 2007;**45**(5): 565-576. DOI: 10.1515/cclm.2007.110

[9] Zandecki M, Genevieve F, Gerard J, Godon A. Spurious counts and spurious results on haematology analysers: A review. Part I: Platelets. International Journal of Laboratory Hematology. 2007;**29**(1):4-20. DOI: 10.1111/j.1365-2257.2006.00870.x

[10] Zandecki M, Genevieve F, Gerard J, Godon A. Spurious counts and spurious results on haematology analysers: A review. Part II: White blood cells, red blood cells, haemoglobin, red cell indices and reticulocytes. International Journal of Laboratory Hematology. 2007;**29**(1):21-41. DOI: 10.1111/j.1365-2257.2006.00871.x

[11] Almagor M, Lavid-Levy O. Effects of blood-collection systems and tubes on hematologic, chemical, and coagulation tests and on plasma hemoglobin. Clinical Chemistry. 2001;**47**(4):794-795

[12] van den Besselaar A, Chantarangkul V, Tripodi A. A comparison of two sodium citrate concentrations in two evacuated blood collection systems for prothrombin time and ISI determination. Thrombosis and Haemostasis. 2000;**84**(4):664-667

[13] Tripodi A, Chantarangkul V, Bressi C, Mannucci PM. How to evaluate the influence of blood collection systems on the international sensitivity index. Protocol applied to two new evacuated tubes and eight coagulometer/thromboplastin combinations. Thrombosis Research. 2002;**108**(1): 85-89. DOI: 10.1016/s0049-3848(02)00383-3

[14] Kratz A, Stanganelli N, Van Cott EM. A comparison of glass and plastic

blood collection tubes for routine and specialized coagulation assays—A comprehensive study. Archives of Pathology & Laboratory Medicine. 2006;**130**(1):39-44

[15] Tetrault GA. Comparison of glass and plastic blood collection tubes. Archives of Pathology & Laboratory Medicine. 2006;**130**(11):1600

[16] Toulon P, Aillaud MF, Arnoux D, Boissier E, Borg JY, Gourmel C. Multicenter evaluation of a bilayer polymer blood collection tube for coagulation testing: Effect on routine hemostasis test results and on plasma levels of coagulation activation markers. Blood Coagulation & Fibrinolysis. 2006; **17**(8):625-631. DOI: 10.1097/01.mbc.0000252595.79282.86

[17] Yavas S, Ayaz S, Kose SK, Ulus F, Ulus AT. Influence of blood collection systems on coagulation tests. Turkish Journal of Hematology. 2012;**29**(4): 367-375. DOI: 10.5505/tjh.2012.59254

[18] D'Angelo G, Villa C. Comparison between siliconized evacuated glass and plastic blood collection tubes for prothrombin time and activated partial thromboplastin time assay in normal patients, patients on oral anticoagulant therapy and patients with unfractioned heparin therapy. International Journal of Laboratory Hematology. 2011;**33**(2): 219-225. DOI: 10.1111/j.1751-553X.2010.01271.x

[19] Xu M, Robbe VA, Jack RM, Rutledge JC. Under-filled blood collection tubes containing K(2)EDTA as anticoagulant are acceptable for automated complete blood counts, white blood cell differential, and reticulocyte count. International Journal of Laboratory Hematology. 2010;**32**(5): 491-497. DOI: 10.1111/j.1751-553X.2009.01211.x

[20] Biljak VR, Bozicevic S, Krhac M, Lovrencic MV. Impact of under-filled blood collection tubes containing K(2) EDTA and K(3)EDTA as anticoagulants on automated complete blood count (CBC) testing. Clinical Chemistry and Laboratory Medicine. 2016;**54**(11):E323-E3E6. DOI: 10.1515/cclm-2016-0169

[21] Lombardi G, Lanteri P, Colombini A, Lippi G, Banfi G. Stability of haematological parameters and its relevance on the athlete's biological passport model. Sports Medicine. 2011; **41**(12):1033-1042

[22] Banfi G, Lundby C, Robach P, Lippi G. Seasonal variations of haematological parameters in athletes. European Journal of Applied Physiology. 2011; **111**(1):9-16. DOI: 10.1007/s00421-010-1641-1

[23] Sanchis-Gomar F, Martinez-Bello VE, Gomez-Cabrera MC, Vina J. Current limitations of the athlete's biological passport use in sports. Clinical Chemistry and Laboratory Medicine. 2011;**49**(9):1413-1415. DOI: 10.1515/cclm.2011.609

[24] Banfi G, Dolci A. Preanalytical phase of sport biochemistry and haematology. The Journal of Sports Medicine and Physical Fitness. 2003; **43**(2):223-230

[25] Ashenden M, Clarke A, Sharpe K, d'Onofrio G, Allbon G, Gore CJ. Preanalytical mixing of whole-blood specimens in the context of the athlete passport. Journal of Clinical Pathology. 2012;**65**(1):8-13. DOI: 10.1136/jclinpath-2011-200269

Chapter 5

In Silico Proteomics EVOO Therapy for Lipid Lowering in the Patients of Diabetes Mellitus

Muhamamd Suhail and Samreen Riaz

Abstract

Diabetes mellitus is a chronic disease caused by inherited and/or acquired deficiency in the production of insulin by the pancreas, or by ineffectiveness of the insulin produced. Diabetes is a life-long disease marked by elevated levels of sugar in the blood. It is considered as the second progressing cause of color blindness and kidney defects throughout the world. There are four times higher chances of getting heart disease and strokes suffering with diabetes than other ailments. Type-2 diabetes has approximately 90% cases in Pakistan and need to be checked out for its therapy. Olive oil could be helpful in diabetes via many of ways, as prolonged inflammation plays a role as an enhancer of diabetes and other diabetes problems. This study reports identification of some of the cheapest ways to lower the cholesterol in the diabetes mellitus in the local population by using advanced technologies. In this work, the main focus of the research was to use olive oil as a therapy, and as a lipid lowering agent, to improve and reduce the chance of blood pressure, lipids, and hence diabetes and cardiovascular diseases.

Keywords: in silico, EVOO, diabetes, proteomic, dyslipidemia

1. In silico proteomics

Bioinformatics is a science of intellectual biology with respect to molecule recognition as it uses informatics principles to sort out the information of molecules on large scale. Applied math, statistics and computer laid down its foundation. In simple words, bioinformatics work as a platform for management of data in molecular biology field. It has many other applications too in molecular biology [1].

Bioinformatics has many applications in different fields of biology like genomics, biotechnology, molecular biology, and many areas of biomedical sciences. For instance, it is used in DNA fingerprinting, agriculture and drug designing approaches [1]. Drug designing is done in silico by the comparative association of ligand protein interactions; leads obtained this way provide innovative medicines. This type of association is the base of synthetic drug designing. The 3-D structure of proteins is studied via pyMOL or VMD softwares. These softwares clearly demonstrate the receptor ligand relationship in drug designing process. The drug developed by this way requires less budget with greater efficiency and less or no harmful side effects. Drug development is done utilizing informatics principles (Lagunin *et al.*, 2014).

Now a days criminal are captured by using bioinformatics tools like phylogenetic analysis provide a proof against the criminal in court of law. In forensic analysis Bayesian statistics are utilized to detect DNA of culprit [2]. Bioinformatics has made of healthcare system modernize with the introduction of personalized drugs etc. [3].

Now it has become possible with the advancement in the knowledge of bioinformatics that genetic defects are pin pointed on time and are tried to be eliminated before their expression. In this way better treatment has become possible in this era (Takeichi *et al.*, 2015). Bioinformatics has revolutionized the agriculture system with the introduction of new and better crops varieties with increased production potential and resistant against crop damaging agents (Katam *et al.*, 2015).

Bioinformatics is not only limited to Allied health fields but it is also play important role in many other fields too as given below;

- Energy sources
- Antibiotic resistance
- Biotechnology
- Bio-weapon manufacturing
- Climate change analysis
- Comparative studies
- Crop betterment
- Development of drought resistant types of plants
- Drug designing
- Evolutionary analysis
- Forensic studies
- Gene therapy
- Genome assembly
- Nutritional quality improvement
- Insect resistance
- Molecular medicine
- Personalized medicine
- Preventative medicine
- Re-sequencing
- Veterinary science
- Waste cleanup process

1.1 *In silico* drug designing

Discovery of new drug is lengthy and expensive process and require lots of mass efforts too. Many of the times the year's hard work in drug discovery is wasted away due to certain factors like toxic effects, poor drug metabolism etc. (Grinter and Zou, 2014).

So with the advancement in the field of informatics, the drug discovery process has been completely changed as drugs are developed in silico with the help of latest Bioinformatics tools and applications. Today many fields use in silico drug designing (**Figure 1**) method like biochemistry, nanotechnology and molecular biology. This method has major advantage on others is its cost effectiveness and quickness (Grinter and Zou, 2014).

Many outstanding tools and softwares are used for drug development by in silico method, those includes;

- PBPK/PD modeling software
- PKUDDS
- Grid computing
- JAVA
- APIS
- Python and Perl
- MEGA
- Clustal W

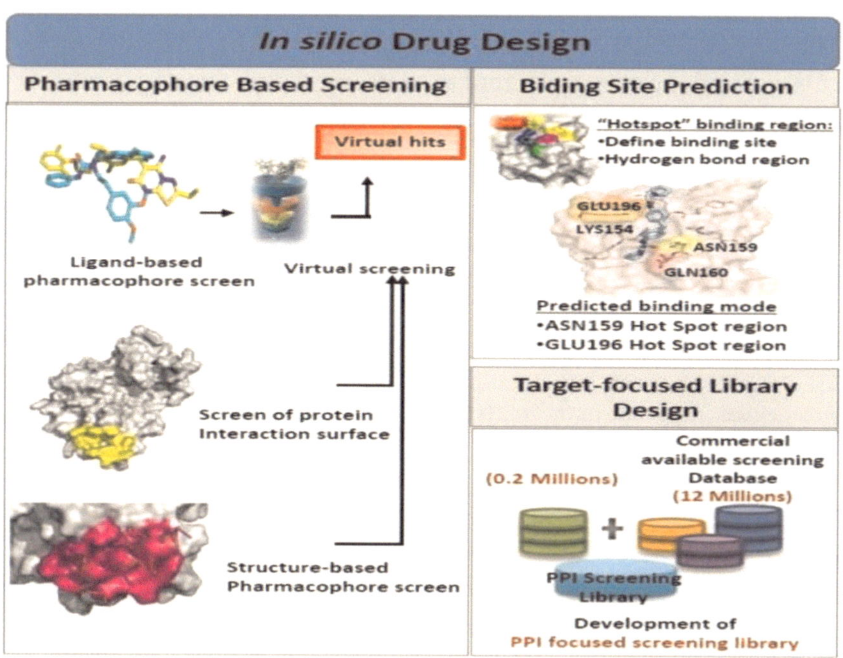

Figure 1.
In silico drug designing.

- Clustal Omega
- JALVIEW
- FATCAT server
- MN convert
- VMD
- Lipinski's rule
- Patch dock &
- AutoDock Vina

This in silico based approach is used in every stage of drug discovery starting from preclinical phase to terminal phase of drug development process [4].

1.2 Phytochemicals

Olea europaea (extra virgin olive oil EVOO) was selected as an anti-diabetic agent as it has many beneficial effects like improvement of hyperglycaemia and lowering of lipid profile of diabetic patients. That is why it was selected and its docking was augmented against certain proteins genetically linked to diabetes.

1.3 Proteins involved in diabetes

There are following proteins which are genetically linked to diabetes, these proteins with their characteristics and available commercial drugs against these protein markers have been shown in **Table 1**.

Protein name	Mol. mass (Da)	Amino acids	Available drugs	Structure
Calcium/calmodulin-dependent protein kinase type 1D	42,914 Da	385	No drug found	*Quaternary structure*
A disintegrin and metalloproteinase with thrombospondin motifs 9	216,491 Da	1935	No drug found	*Quaternary structure*
Calpain-10	74,952 Da	672	No drug found	*Quaternary structure*
Cell division cycle protein 123 homolog	39,135 Da	336	No drug found	*Quaternary structure*
Threonylcarbamoyladenosine tRNA methylthiotransferase	65,111 Da	579	No drug found	*Quaternary structure*
Hematopoietically expressed homeobox protein HHEX	30,022 Da	270	No drug found	*Quaternary structure*
Insulin-degrading enzyme	117,968 Da	1019	6bK inhibitor	*Quaternary structure* Homodimer
Insulin-like growth factor 2 mRNA-binding protein 2	66,121 Da	599	No drug found	*Quaternary structure*

Protein name	Mol. mass (Da)	Amino acids	Available drugs	Structure
Juxtaposed with another zinc finger protein 1	27,079 Da	243	No drug found	*Quaternary structure*
ATP-sensitive inward rectifier potassium channel 11	43,541 Da	390	Sulfonylurea	*Quaternary structure*
Potassium voltage-gated channel subfamily KQT member 1	74,699 Da	676	Incretin	*Quaternary structure Tetramer*
Leucine-rich repeat-containing G-protein coupled receptor 5	99,998 Da	907	No drug found	*Quaternary structure*
Melatonin receptor type 1B	40,188 Da	362	Melatonin supplements, DPP4 inhibitors	*Quaternary structure*
Peroxisome proliferator-activated receptor gamma	57,620 Da	505	Thiazolidinediones, pioglitazone	*Quaternary structure*
Transcription factor 7-like 2	67,919 Da	619	Sulfonylurea, linagliptin, incretin	*Quaternary structure*
Thyroid adenoma-associated protein	219,607 Da	1953	No drug found	*Quaternary structure*
Tetraspanin-8	26,044 Da	237	No drug found	*Quaternary structure*
Wolframin	100,292 Da	890	Incretin	*Quaternary structure*

Table 1.
Proteins involved in diabetes along with their properties, structure and available drugs has been demonstrated.

1.4 Significance of *in silico*, drug designing

In silico medical practitioner to select best suitable drug for the treatment of disease. The drug designed by this way prove to be less toxic, fewer or no side effect and more capable than any other drug available. In silico drug designing has application in all the phases of new drug development process. All these in silico analysis are done with the help of different tools and application softwares of bioinformatics (Schmidt *et al.*, 2014; Tian *et al.*, 2015).

1.5 Aims of study

In silico bioinformatics analysis were used to screen and check the therapeutic potential of *Olea europaea* to treat type two diabetes mellitus. For this purpose certain proteins responsible for diabetes were selected and in silico analysis were done to check the reversion effect of *Olea europaea* against these diseases causing proteins expression. The purpose of the study was to screen and isolate potential therapeutic agent present in the *Olea europaea* which could be the best treatment option of diabetes.

1.6 Future prospective

The *in silico* drug development could be meaningful in the certain aspects:

- Reduction of drug invention time and cost
- Harmless, effective and target specific drug development
- To improvement in result analysis
- Develop efficient bioinformatics ways and algorithms for
- Lead identification
- Lead optimization
- Target identification
- Target validation

2. Diabetes mellitus

Diabetes is a persistent ailment which happens because of the raised blood glucose due to the lack of the body' ability to produce insulin secretion or sufficient insulin secretions. Insulin secretion is obtained from the pancreas gland in the body and is prerequisite for the transport of glucose from bloodstream to the living cells where it is utilized to obtain energy for the body. Deficit of insulin or impotence of cells to response to insulin results in high level of blood glycaemia or hyperglycaemia which is distinctive feature of diabetes. Hyperglycaemia, if remained uncontrolled over long period, can cause severe damage to different body parts leading to development of disabling or many life-threatening health issues like cardiovascular disorders, neuropathy, nephropathy and eye diseases leading to retinopathy and blindness. Proper management of diabetes helps prevent or delay those complications (Atlas 2017 intro 1).

The term diabetes was initiated from Greek word "syphon" or "mellitus" means Honeyed proposed by Rollo in late eighteenth century [5]. Matthew Dobson for the first time reported the hyperglycemic nature of diabetes in his article during 1776; he observed elevated glucose level in urine and plasma of patients. Holt et al. [6] reported the raised level of ketone bodies in diabetes.

Million people die of diabetes each year, mostly from cardiovascular disease exacerbated by lipid disorders and other complications. Identification of new treatment strategies using olive oil for lipid lowering in diabetes through proteomic analysis was needed. With the increase in the world population the proportion of diabetes is also increasing with great speed so, the more health care budget is needed to cope with this problem. As without proper primary strategies and care it will take the face of epidemic which will kill millions of lives in the world. With this speed it is possible that diabetes would become the number one killer of the world within next coming 25 years. Abrupt actions are required to root out the diabetes from the world.

Olive oil supplementation with daily food intake might play a protective role to reduce lipid profile in diabetes due to the beneficial effect of olive oil exerted by its antioxidant constituents.

In most of the cases when the disease is detected it is already too late to be cured. The percentage of fatalities in patients is therefore very high. Not only there is a vast number of precious lives are lost but the treatment and maintenance of such patients is a heavy financial burden on the individual families and the national resources. It is therefore necessary that cheap and quick methods are made available which can be used to screen population so that the diseases can be detected at a very early stage. Research and development work based on the samples from our population groups is essential because findings have been reported to be dependent on race variations, and socio-economic and other factors. Thus the studies reported on the basis of population groups in other parts of the world may not be valid at least in some cases.

This study reports identification of some of the cheapest way to lower the cholesterol profile in the diabetes mellitus in the local population by using advance technologies. In this work, the prime focus was to the use of olive oil as therapy for cholesterol lowering agent to improve and reducing the chance of blood pressure, lipids and hence diabetes and cardiovascular disease will be discussed. The findings from present research work would prove better treatment options for diabetic patients by introduction and characterization of new therapy methods.

DM is one out of many lethal melodies. It is ranked fourth deadliest malady, and the rate of prevalence is increasing sharply. Olive oil a liquid fat captured from olives tree crop cultivating in the Mediterranean Basin. Complete olives are pressed to gain olive oil.

Results generated from this research will help devise for the novel diagnostic procedures for early and control of diabetes complications in our population. These findings will provide a base in planning better treatment strategies for diabetic patients by suggested dose of olive oil nutritional supplement. The research work proposed to be undertaken under this project shall elucidate the roles of olive oil in diabetes mellitus in our population. Therapy for the diabetes to be investigated should lead to not only saving lot of precious lives but also ensure saving of huge numbers of man-hours and colossal amount financial resources spent on treatment and patient maintenance. The other financial gains accruing from this project is the application of the project results in developing therapy methods locally for routine use in our clinical laboratories.

2.1 Classification

Diabetes has been grouped into two types since 600 BC but the clear classification into two types was done in nineteenth century. Diabetes of young before ketoacidosis and diabetes of obese people, later these two types were named as MODY by WHO. By the discovery of human leukocyte antigen (HLA) and antibodies in 1970, it was become clear that all the patients on insulin treatment were suffering with autoimmune disorder [6]. WHO revised the classification into two types of diabetes in 1980 and 1985 based on the clinical findings, which were insulin-dependent diabetes mellitus (IDDM) and the noninsulin-dependent diabetes mellitus (NIDDM). Classification of 1999 was accepted and revised with small changes and diabetes was grouped into (T1DM), (T2DM), gestational diabetes mellitus, and some other types [7].

For the diagnosis of diabetes, the diagnostic criterion has been adopted and revised over and again by world health organization (WHO). The current criteria states that diabetes is diagnosed by the elevated level of glucose in the bloodstream of diabetic patients (**Table 2**).

2.2 Type 1 diabetes mellitus

T1DM is an autoimmune ailment caused when the body's immune system attack on the Beta-cells in the islets of pancreas gland. As a conclusion body produce very

DIABETES should be diagnosed if ONE OR MORE of the following criteria are met	IMPAIRED GLUCOSE TORELANCE (IGT) should be diagnosed if BOTH of the following criteria are met	IMPAIRED FASTING GLUCOSE (IFG) should be diagnosed if BOTH of the following criteria are met
Fasting plasma glucose ≥7.0 mmol/L (126 mg/dL)	Fasting plasma glucose <7.0 mmol/L (126 mg/dL)	Fasting plasma glucose 6.1-6.9 mmol/L (110 to 125 mg/ dL)
or	and	and
Two-hour plasma glucose ≥11.1 mmol/L (200 mg/dL) following a 75g oral glucose load	Two-hour plasma glucose ≥7.8 <11.1mmol/L (≥140 to <200 mg/dL) following a 75g oral glucose load	Two-hour plasma glucose <7.8mmol/L (140mg/dL) following a 75g oral glucose load
or		
A random glucose > 11.1 mmol/L (200 mg/ dL) or HbA1c ≥ 48 mmol/mol (equivalent to 6.5%)		

Table 2.
Diabetes diagnostic criteria.

little or none insulin with a relative or absolute deficiency of insulin in the body. The reason of this deadliest process are not completely understood but it is thought that genetic susceptibility and environmental factors like viral infection, toxins and some dietary factors could be the cause of the problem (Chapt1 type1 10). Type 1 diabetes occurs mostly in children and adolescents. It is managed with proper daily injections of insulin otherwise the survival of the patient would be impossible. Type 1 is diagnosed by raised level of glucose and with the specific symptoms (**Figure 2**). Type 1 diabetes incidence rate in Karachi, the largest city of Pakistan is 1.02/100,000 in comparison to type 2 diabetes mellitus [8].

2.3 Type 2 diabetes mellitus

Type 2 diabetes is commonest type of diabetes throughout the world accounting around 90% of all the diabetic patients. It is occurred due to insufficient production of insulin or insulin resistance. During insulin resistance insulin become ineffective as a result, body produce excessive insulin initially to control the elevated level of

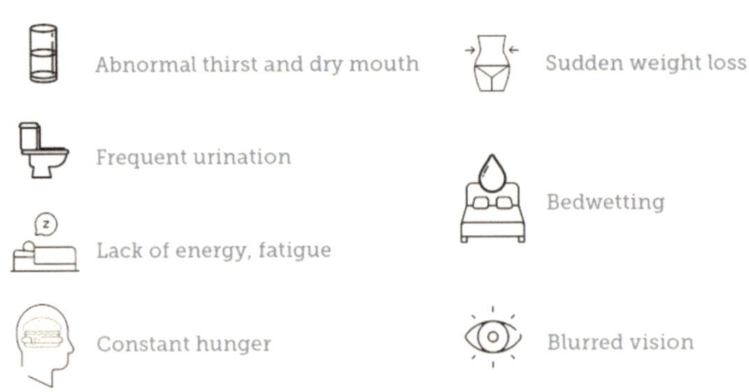

Figure 2.
The symptoms of type 1 diabetes.

glucose but as the time passes productive ability of insulin producing cell is lost ultimately a state of inadequate insulin production is developed. Usually type 2 diabetes occurs in older adults but it is increasingly seen in children, adolescent and young adults due to physical inactivity, poor diet increasing level of obesity (13–15). Type 2 diabetes alarming incidence in Pakistan is associated with poor diet etc. [9]. The estimated prevalence of Pakistan ranges from 7.6 to 11% in 2011 [10]. The symptoms of T2DM are like T1DM (**Figures 3** and **4**).

2.4 Gestational diabetes mellitus (GDM)

GDM is a non-pathophysiological condition and one out of seven new born has gestational diabetes according to [11]. GDM caused by glucose intolerance which was first diagnosed in pregnancy period and may remain after pregnancy (Gavin III *et al.*, 1997). The development of undiagnosed asymptomatic T2DM and T1DM during pregnancy is referred as GDM [12]. The occurrence of GDM is reported during the last 6 months of the pregnancy while the development of diabetes in women during first trimester is termed as T2DM [13].

2.5 Type 1.5 diabetes

Latent Autoimmune Diabetes in Adults (LADA) non-officially called as Type 1.5 diabetes. The term 1.5 indicates that disease is a type 1 diabetes mellitus but it shares

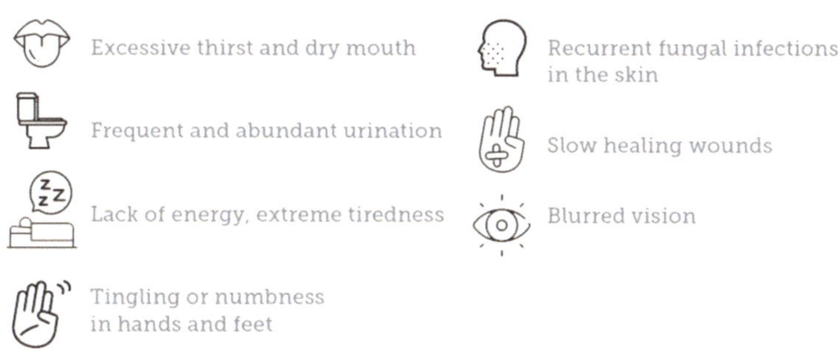

Figure 3.
The symptoms of typ. 2 diabetes.

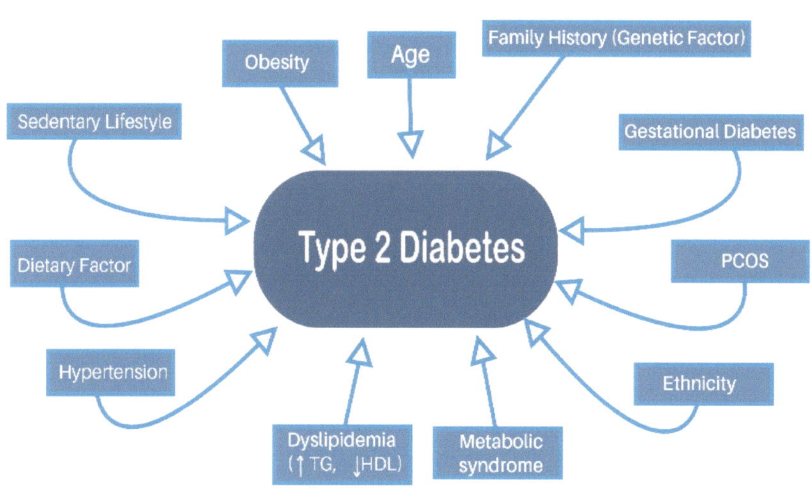

Figure 4.
Different risk factors of typ. 2 diabetes mellitus (T2DM).

some feature with type 2 diabetes mellitus that is the reason it is called as Type 1.5 diabetes means in between T1DM and T2DM. It is also known as Type 3 diabetes. Type 1.5 diabetes is diagnosed during adulthood and has slow onset similar to T2DM however, it is type 1 diabetes mellitus and will need insulin therapy in future. Around 15–20% patients with type 2 diabetes are actually suffering with type 1 diabetes. They are treated with oral medication initially but will need insulin therapy after some period of oral treatment. Type 1.5 diabetic patients do not have standard symptoms of type 2 diabetes and have reduced risk of heart problem if their blood sugar is kept in control (Diabetes, UK).

2.6 Other specific types of diabetes

Some other types of diabetes occur due to the genetic malfunctions, impaired secretory activity of pancreas and over production of certain hormones. For instance MODY situation affects certain genes which are responsible for beta-cells development, functioning and regulation of glucose. Likewise excessive production certain growth hormones, cortisol, epinephrine and glucagon enhance the glucose production in liver cells and increase insulin resistance in muscle cells resulting in the progression of diabetes mellitus [14]. Cystic fibrosis–related diabetes (CFRD) occurred in cystic fibrosis patients due to insulin resistance and genetic dysfunction of beta-cell linked with inflammation and infection [15].

2.7 Epidemiology

With every passing decade individual with diabetic ailment are increasing tremendously, currently there are 425 million people suffering with diabetes and 352 million people have falsified glucose tolerance (IGT). IGT is also named as impaired fasting glucose (IFG) or "Pre-Diabetes". The individual with IFG has been at peak

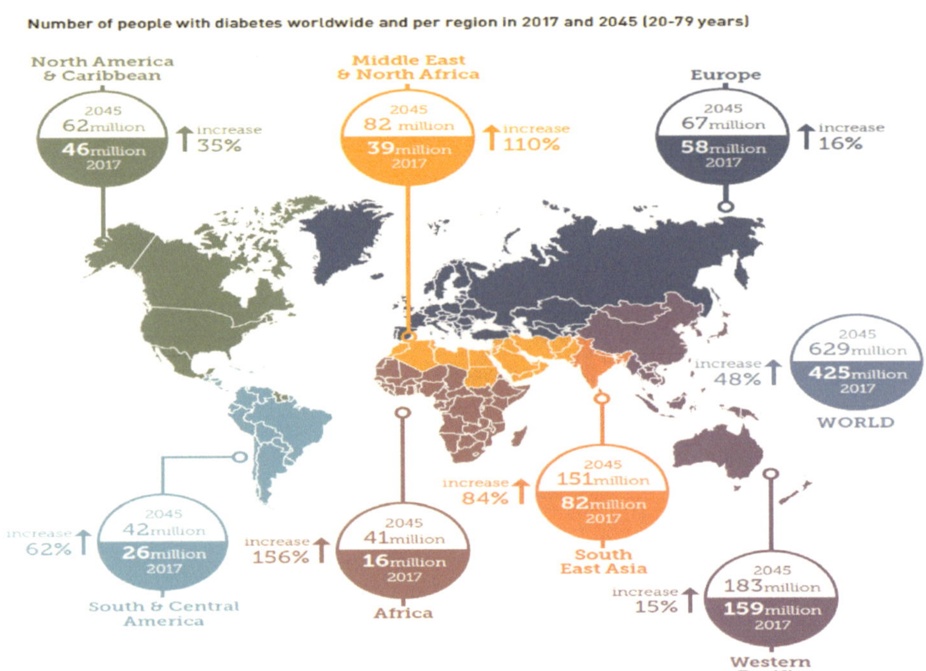

Figure 5.
Worldwide distribution of diabetes. Adapted from: IDF diabetes (IDF 2017).

risk of diabetes in near future. According to IDF, 1 in 11 people has diabetes till now and it is projected to be 1 in 10 till 2040 (IDF 2017).

Pakistan is among the top 10 most affected diabetic countries of the world and is projected to be at 8th position until 2045 (20–79 years of age. Other major countries on the top of the list are China, India, United States, Brazil, Mexico, Indonesia, Russian federation, Egypt, Germany and Pakistan respectively. According to latest estimates Pakistan has 7.2 million population of diabetes and projected to be 16.1 million by 2045 (Atlas 2017). Previous studies reported 5.2 million diabetic population of Pakistan in 2000 and projected to be 13.9 million till 2030 [16]. The worldwide prevalence of diabetes is shown in **Figure 5**. In urban areas the ratio of diabetes is 22.04% and in rural areas it is 17.15%. In Punjab, diabetes in female is 19.3% and in male it is 16.6%. In KPK it is 11.1% in both male and female in Balochistan it is 10.8% in both [17, 18]. But in Sindh female proportion is 11.7% and in male is 16.2% (Hussain and Ali, 2017) **Figure 6** demonstrate the distribution of diabetes in all the provinces of Pakistan (**Figure 7**) [19–39].

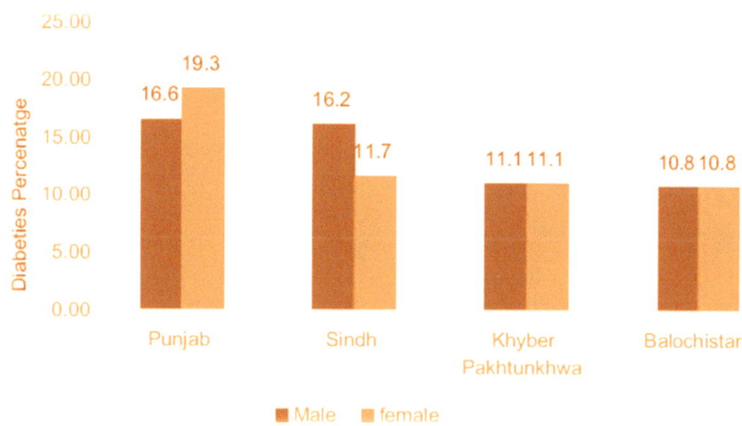

Figure 6.
Distribution of diabetes in different provinces of Pakistan [40].

2017			2045		
Rank	Country/territory	Number of people with diabetes	Rank	Country/ territory	Number of people with diabetes
1	China	114.4 million (104.1-146.3)	1	India	134.3 million (103.4-165.2)
2	India	72.9 million (55.5-90.2)	2	China	119.8 million (86.3-149.7)
3	United States	30.2 million (28.8-31.8)	3	United States	35.6million (33.9-37.9)
4	Brazil	12.5 million (11.4-13.5)	4	Mexico	21.8 million (11.0-26.2)
5	Mexico	12.0 million (6.0-14.3)	5	Brazil	20.3 million (18.6-22.1)
6	Indonesia	10.3 million (8.9-11.1)	6	Egypt	16.7million (9.0-19.1)
7	Russian Federation	8.5 million (6.7-11.0)	7	Indonesia	16.7million (14.6-18.2)
8	Egypt	8.2million (4.4-9.4)	8	Pakistan	16.1 million (11.5-23.2)
9	Germany	7.5 million (6.1-8.3)	9	Bangladesh	13.7 million (11.3-18.6)
10	Pakistan	7.5 million (5.3-10.9)	10	Turkey	11.2 million (10.1-13.3)

Figure 7.
List of top 10 countries having diabetes (IDF Atlas 2017).

2.8 Prevention of diabetes

As diabetes is multifactorial gene disorder, as genetic information and environmental effects both play a role in this disease. Type 1 diabetes is comparatively less affected with environmental factors but type 2 does. So, it is difficult to stop the onset of type 1 diabetes, but for type 2 diabetes there are many known preventive measures. These preventive measures include maintaining normal body weight, use of healthful diet and regular physical exercise. Changing in eating habits are known to reduce the risk of diabetes, such as taking whole grain diet, using diet with good fats (nut, fish) and reducing use of saturated fats (red meat). But above all, limit use of sweet edibles is the most important and effective preventive measure as well as in controlling the complication of diabetes. In addition, if an individual quits smoking, the risk of onset of diabetes reduces many folds (**Figure** 7) [41].

3. Extra virgin olive oil (EVOO)

EVOO is one of the most important components of Mediterranean diet having many health benefits. At first, excess of (MUFA) especially oleic acid, was thought

Figure 8.
Representing health benefits of extra virgin olive oil.

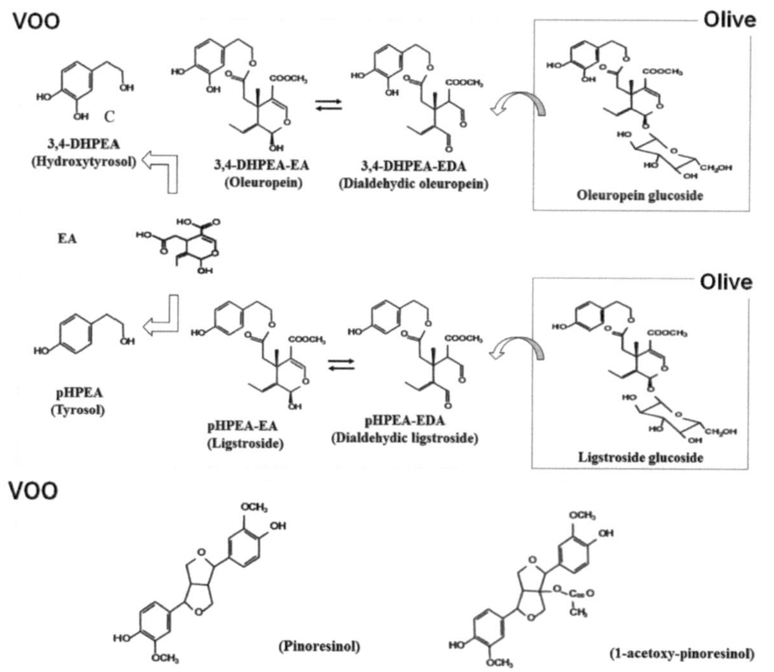

Figure 9.
Chemical structure of EVOO.

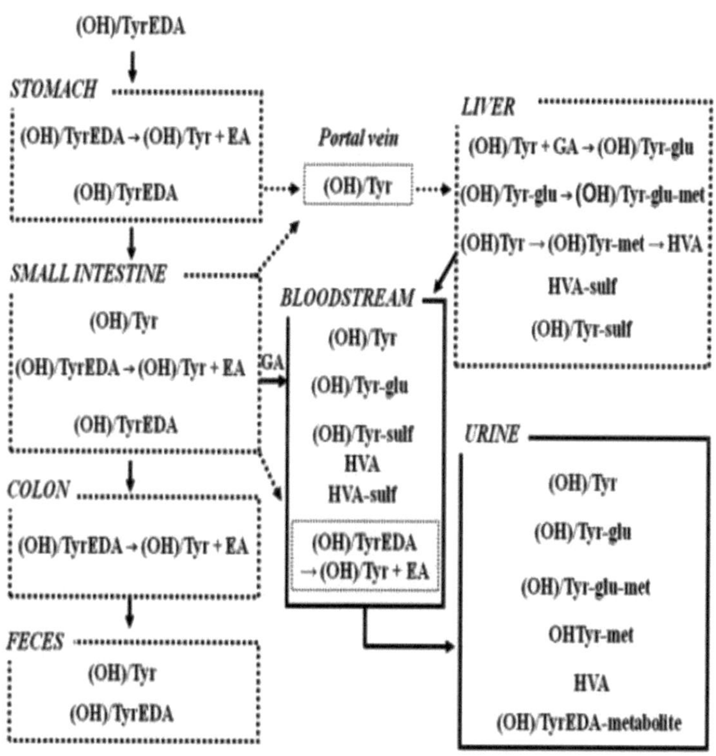

Figure 10.
Metabolism of EVOO.

as the major beneficial constituent of virgin olive oil. Most important health benefit of EVOO (**Figure 8**) is improved cardiovascular heal, reduction of inflammation and oxidative stress. It is used in Mediterranean diet since many decades. Studies suggested that Mediterranean food specimen along with EVOO lowers the risk of heart disorders as compared to controls. Other studies reported that EVOO improve hyperglycemia and lipid level in patients with dysfunctional fasting glucose (Roberto *et al.*, 2016). Almost 200 'minor particles' in the un-saponifiable portion of olive oil comprise about 2% of the total weight of EVOO [42–63].

3.1 Nutritional effects of EVOO

One tablespoon of EVOO contains 14 g of total fat, 2 g of saturated fats and has no fiber, sugar, cholesterol etc. Vitamins E, K are also present in it also has phytosterols, a which is structurally similar to cholesterol because it hinders cholesterol absorption from food and lower LDL values associated with heart disease. Major EVOO polyphenols include secoiridoids, phenols, and lignans. The chemical structural formula of EVOO is shown in **Figure 9** and metabolism of EVOO is shown in **Figure 10**.

3.2 Effects on cholesterol level

There is no cholesterol in the EVOO while other animal obtained fats contain different amount of cholesterol. Recently, the European Food Safety Authority (EFSA) has reported beneficial effect of EVOO on LDL oxidation process due to its phenolic (**Figure 10**).

4. Justification

Diabetes mellitus is one of the most lethal melodies of the world and ranked fourth in deadliest diseases. It is metabolic defect that results in hyperglycaemia. The distribution of diabetes is spreading with great speed. The diabetic mass of Pakistan is among the top 10 diabetic populations of the world and is projected to be at 8th position among most prevailing diabetic countries by 2045 (IDF 2017). With this rate 1 out of 4 persons in Pakistan is diabetic. It is not a single etiological condition but it is a multifactorial ailment that is affecting Pakistan population severely. Diabetes has no cure at the moment only preventive and supportive therapy is given to improve the living of people by means of oral medication and insulin injections [63–72].

Million people die of diabetes each year, mostly from cardiovascular disease exacerbated by lipid disorders and other complications.

Olive oil use has may a protective role to reduce the lipid profile in diabetes due to the beneficial effect of its antioxidant level. Since oral medications and insulin are the best possible treatment option but still these have some side effects like drugs toxicity, renal failure and inflammation and infection at the point of injection of insulin to minimize the effect and to obtain a better alternative treatment option which should be natural, harmless, cheap and could prove a permanent cure of diabetes we chose EVOO as a therapy of the type 2 diabetes.

Since there are some recent studies available which support the use of EVOO as a lipid and glucose lowering agent in type 2 diabetes mellitus. EVOO improves the post prandial glycaemia by incretin like mechanism in which up-regulation of incretin took place in this way lipid and glucose levels are reduce by EVOO. EVOO also reduce oxidative stress, inflammation lower lipid level which is the hallmark of the diabetes that is why EVOO is compound of choice for the therapy of diabetes. The phenols in the EVOO are the major components responsible for these beneficial effects. Due of the fact that the EVOO therapy for the treatment of diabetes has never been studied before in the diabetic population of Pakistan we step forward and to use this natural compound as a therapy for treatment of type 2 diabetes [72–80].

There is no study available in Pakistan that support the therapeutic effect of EVOO in type 2 diabetes so got the idea to prove the therapeutic value of EVOO in the Diabetic population of Pakistan, for this we selected the Lahore region that is one of the populated city of Pakistan.

Author details

Muhamamd Suhail and Samreen Riaz*
Department of Microbiology and Molecular Genetics, University of the Punjab, Lahore, Pakistan

*Address all correspondence to: samreen.mmg@pu.edu.pk

References

[1] Egea JA, Henriques D, Cokelaer T, Villaverde AF, MacNamara A, Danciu D, et al. MEIGO: An open-source software suite based on metaheuristics for global optimization in. Systems Biology and Bioinformatics. Bioinformatics. 2014;**15**:136-138

[2] Van Neste C, Gansemans Y, De Coninck D, Van Hoofstat D, Van Criekinge W, Deforce D, et al. Forensic massively parallel sequencing data analysis tool: Implementation of MyFLq as a standalone web-and Illumina BaseSpace®-application. Forensic Science International: Genetics. 2015; **15**:2-7

[3] Shameer K, Badgeley MA, Miotto R, Glicksberg BS, Morgan JW, Dudley JT. Translational bioinformatics in the era of real-time biomedical, health care and wellness data streams. Briefings in Bioinformatics. 2017;**18**:105-124

[4] Sliwoski G, Kothiwale S, Meiler J, Lowe EW. Computational methods in drug discovery. Pharmacological Reviews. 2014;**66**:334-395

[5] Waly MI, Ali A, Essa MM. International Journal of Biological & Medical Research. 2010;**1**(4):202-205

[6] Holt RIG, Cockram C, Flyvbjerg A, Goldstein BJ. Textbook of Diabetes. Vol. 3. John Wiley & Sons; 2011

[7] Kuzuya T, Matsuda A. Classification of diabetes on the basis of etiologies versus degree of insulin deficiency. Diabetes Care. 1997;**20**(2):219

[8] Hakeem R, Fawwad A. Diabetes in Pakistan: Epidemiology, determinants and prevention. Journal of Diabetology. 2010;**3**(4)

[9] Zafar J, Bhatti F, Akhtar N, Rasheed U, Bashir R, Humayun S, et al. Prevalence and risk factors for diabetes mellitus in a selected urban population of a city in Punjab. Journal of the Pakistan Medical Association. 2011; **61**(1):40

[10] Ansari RM, Dixon JB, Coles J. Type 2 diabetes: Challenges to health care system of Pakistan. International Journal of Diabetes Research. 2015. DOI: 10.5923/j.diabetes.20150401.02

[11] Lau DCW. Diabetes in the elderly: A silent global tsunami. Canadian Journal of Diabetes. 2016;**40**(1):2-3. DOI: 10.1016/j.jcjd.2015.12.003

[12] Baynest HW. Classification, pathophysiology, diagnosis and management of diabetes mellitus. Journal of Diabetes & Metabolism. 2015: 2015

[13] American Diabetes, Association. 2. Classification and diagnosis of diabetes. Diabetes Care. 2015;**38**(Suppl):S8-S16. DOI: 10.2337/dc15-S005

[14] Triplitt, Curtis, Solis-Herrera, Carolina, Reasner, Charles, DeFronzo, Ralph A, & Cersosimo, Eugenio. (2015). Classification of Diabetes Mellitus.

[15] American Diabetes, Association. 2. Classification and diagnosis of diabetes. Diabetes Care. 2016;**39**(Supp. 1):S13-S22. DOI: 10.2337/dc16-S005

[16] Rahim M, Rahim M, Rahim M, Sharafat S, Shaikh Z. Maternal and paternal transmission of diabetes: Influence of nutritional factors. Journal of Diabetes and Metabolism. 2015;**6**(3): 504. DOI: 10.4172/2155-6156.1000 504. Page 2 of 7 Volume 6• Issue 3• 1000504 J Diabetes Metab Issn: 2155-6156 Jdm, an Open Access Journal Figure 1: Percentages of Age Frequency (Vol. 2, pp. 3): Figure

[17] Afzal S, Brøndum-Jacobsen P, Bojesen SE, Nordestgaard BG. Vitamin

D concentration, obesity, and risk of diabetes: A Mendelian randomisation study. The Lancet Diabetes & Endocrinology. 2014;**2**(4):298-306

[18] American Diabetes, Association. Standards of medical care in diabetes—2016 abridged for primary care providers. Clinical Diabetes. 2016;**34**(1): 3-21. DOI: 10.2337/diaclin.34.1.3

[19] Anguizola J, Matsuda R, Barnaby OS, Hoy KS, Wa C, DeBolt E, et al. Review: Glycation of human serum albumin. Clinica Chimica Acta. 2013; **425**:64-76. DOI: 10.1016/j.cca.2013.07.013

[20] IDF Diabetes Atlas. IDF Diabetes Atlas, 7th ed. International Diabetes Federation; 2015

[21] Babu J, Malkiwodeyar PK, Tarikere S, Cholenahalli NM. Myriad cardiac manifestation of hyperhomocysteinemia. Cardiology Research. 2015;**6**(2):263-265

[22] Barnabé A, Aléssio ACM, Bittar LF, Mazetto d M, Bruna B, Angélica M, et al. Folate, vitamin B12 and homocysteine status in the post-folic acid fortification era in different subgroups of the Brazilian population attended to at a public health care center. Nutrition Journal. 2015;**14**(1):1

[23] Beaudin AE, Stover PJ. Folate-mediated one-carbon metabolism and neural tube defects: Balancing genome synthesis and gene expression. Birth Defects Research Part C: Embryo Today: Reviews. 2007;**81**(3):183-203. DOI: 10.1002/bdrc.20100

[24] Berg AH, Drechsler C, Wenger J, Buccafusca R, Hod T, Kalim S, et al. Carbamylation of serum albumin as a risk factor for mortality in patients with kidney failure. Science Translational Medicine. 2013;**5**(175):175ra129-175ra129

[25] Bonetti F, Brombo G, Magon S, Zuliani G. Cognitive status according to homocysteine and B-group vitamins in elderly adults. Journal of the American Geriatrics Society. 2015;**63**(6):1158-1163

[26] Bradford MM. A rapid and sensitive method for the quantitation of microgram quantities of protein utilizing the principle of protein-dye binding. Analytical Biochemistry. 1976; **72**(1–2):248-254

[27] Bryan J. Folic acid: From a rescue treatment for methotrexate patients to landmark drug. Spinal Surgery. 2015; **14**:03

[28] Bukhari SA, Javed S, Ali M, Shahzadi A, Rehman M. Serum haematological and biochemical indices of oxidative stress and their relationship with DNA damage and homocysteine in Pakistani type II diabetic patients. Pakistan Journal of Pharmaceutical Sciences. 2015;**28**(3):881-889

[29] Chan YM, Bailey R, O'Connor DL. Folate. Advances in Nutrition. 2013; **4**(1):123-125. DOI: 10.3945/an.112.003392

[30] Cheung ENM. Synthesis and Application of Polymer Stabilized Lanthanide Fluoride Nanoparticles. University of Toronto; 2010

[31] Ericson U, Sonestedt E, Gullberg B, Hellstrand S, Hindy G, Wirfält E, et al. High intakes of protein and processed meat associate with increased incidence of type 2 diabetes. British Journal of Nutrition. 2013;**109**(06):1143-1153

[32] Evans JL, Goldfine ID. A new road for treating the vascular complications of diabetes: So let's step on the gas. Diabetes. 2016;**65**(2):346-348

[33] Feng Y, Shan M-Q, Bo L, Zhang X-Y, Hu J. Association of homocysteine with type 1 diabetes mellitus: A meta-analysis. International Journal of Clinical and Experimental Medicine. 2015;**8**(8):12529-12538

[34] Furusyo N, Hayashi J. Glycated albumin and diabetes mellitus. Biochimica et Biophysica Acta (BBA)-General Subjects. 2013;**1830**(12): 5509-5514

[35] Gornall AG, Bardawill CJ, David MM. Determination of serum proteins by means of the biuret reaction. The Journal of Biological Chemistry. 1949; **177**(2):751-766

[36] Gray N, Picone G, Sloan F, Yashkin A. Relation between BMI and diabetes mellitus and its complications among us older adults. Southern Medical Journal. 2015;**108**(1):29-36

[37] Group, National Diabetes Data. Classification and diagnosis of diabetes mellitus and other categories of glucose intolerance. Diabetes. 1979;**28**(12): 1039-1057

[38] Iqbal MP. Hyperhomocysteinemia and coronary artery disease in Pakistan. The Journal of the Pakistan Medical Association. 2006;**56**: 282-285

[39] Iqbal MP, Ishaq M, Kazmi KA, Yousuf FA, Mehboobali N, Ali SA, et al. Role of vitamins B6, B12 and folic acid on hyperhomocysteinemia in a Pakistani population of patients with acute myocardial infarction. Nutrition, Metabolism and Cardiovascular Diseases. 2005;**15**(2): 100-108

[40] Iqbal P, Lindblad BS, Mehboobali N, Yusuf FA, Khan AH, Iqbal SP. Folic acid and vitamin B6 deficiencies related hyperhomocysteinemia in apparently healthy Pakistani adults; is mass micronutrient supplementation indicated in this population. Journal of the College of Physicians and Surgeons Pakistan. 2009;**19**(5):308

[41] James R GIII, Alberti KGMM, Davidson, Mayer B, DeFronzo RA. Report of the expert committee on the diagnosis and classification of diabetes mellitus. Diabetes Care. 1997; **20**(7):1183

[42] FAO & World Health Organization Joint. Vitamin and Mineral Requirements in Human Nutrition; 2005

[43] Kruger NJ. The Bradford method for protein quantitation. In: Basic Protein and Peptide Protocols. 1994. pp. 9-15

[44] Kumar J, Ingelsson E, Lind L, Fall T. No evidence of a causal relationship between plasma homocysteine and type 2 diabetes: A Mendelian randomization study. Frontiers in Cardiovascular Medicine. 2015;**2**:11. DOI: 10.3389/fcvm.2015.00011

[45] Kurien BT, Scofield RH. Extraction of proteins from gels: A brief review. In: Biji Kurien T, Scofield HR, editors. Protein Electrophoresis: Methods and Protocols. Totowa, NJ: Humana Press; 2012. pp. 403-405

[46] Laanpere M, Altmäe S, Stavreus-Evers A, Nilsson TK, Yngve A, Salumets A. Folate-mediated one-carbon metabolism and its effect on female fertility and pregnancy viability. Nutrition Reviews. 2010;**68**(2):99-113

[47] Loghmani E. Diabetes mellitis: Type 1 and type 2. Diabetelogia. 2005; **27**:998-1010

[48] Mathis D, Vence L, Benoist C. B-cell death during progression to diabetes. Nature. 2001;**414**(6865):792-798

[49] McRae MP. High-dose folic acid supplementation effects on endothelial function and blood pressure in hypertensive patients: A meta-analysis of randomized controlled clinical trials. Journal of Chiropractic Medicine. 2009; **8**(1):15-24. DOI: 10.1016/j.jcm.2008.09.001

[50] WHO Diabetes Mellitus Group. Report of a WHO Study. Technical

Report Series No. 727. Geneva: World Health Organization; 1985. pp. 199-205

[51] Moore EM, Ames D, Mander AG, Carne RP, Brodaty H, Woodward MC, et al. Among vitamin B12 deficient older people, high folate levels are associated with worse cognitive function: Combined data from three cohorts. Journal of Alzheimer's Disease. 2014; **39**(3):661-668

[52] Nasri H, Baradaran A. Association of serum homocysteine with anemia in maintenance hemodialysis patients. Pakistan Journal of Nutrition. 2005; **4**(6):414-417

[53] Nazki FH, Sameer AS, Ganaie BA. Folate: Metabolism, genes, polymorphisms and the associated diseases. Gene. 2014;**533**(1):11-20. DOI: 10.1016/j.gene.2013.09.063

[54] NCBI. Folr1: Human folate receptor 1 (report). (Gene ID: 2348). 2016. Retrieved from: NCBI http://www.ncbi.nlm.nih.gov/gene/2348 [29 May 2016]

[55] National Institute of Health (NIH). Folate: Dietary supplement fact sheet. 20 April, 2016. Retrieved from: https://ods.od.nih.gov/factsheets/Folate-Health Professional/#h2 [26 May 2016]

[56] National Institute of Health (NIH). Folr1 Folate Receptor 1. 5 May, 2016. Retrieved from: https://ghr.nlm.nih.gov/gene/FOLR1 [29 May 2016]

[57] O'Neill RJ, Vrana PB, Rosenfeld CS. Maternal methyl supplemented diets and effects on offspring health. Frontiers in Genetics. 2014;**5**:289. DOI: 10.3389/fgene.2014.00289

[58] Orchard TJ, Temprosa M, Barrett-Connor E, Fowler SE, Goldberg RB, Mather KJ, et al. Long-term effects of the diabetes prevention program interventions on cardiovascular risk factors: A report from the Dpp outcomes study. Diabetic Medicine. 2013;**30**(1):46-55

[59] World Health Organization. Definition and Diagnosis of Diabetes Mellitus and Intermediate Hyperglycemia. 2006. Importance and Transferability; 2010

[60] Pubchem. Folic acid. 2016. Available from: PubChem Compound Database. Retrieved from: National Center for Biotechnology Information https://pubchem.ncbi.nlm.nih.gov/compound/91552 [07 May 2016]

[61] Queralt VC. The Role of Motivation, Self-Efficacy, Illness Representations and Family Responsibility in Relation to Diabetes Outcomes: Perceptions of Adolescents with Type 1 Diabetes and their Parents. University of Birmingham; 2010

[62] Riaz S. Evaluation and analysis of human folate levels in Pakistani diabetic population. International Journal of Scientific & Engineering Research. 2014;**5**(12):1572

[63] Riaz S. Study of protein biomarkers of diabetes mellitus type 2 and therapy with vitamin B1. Journal Diabetes Research. 2015;**2015**:150176. DOI: 10.1155/2015/150176

[64] Riaz S, Alam SS, Raza M, Hasnain S, Akhtar MW. Obesity as risk factor and study of obesity related proteins in diabetes mellitus. African Journal of Biotechnology. 2009;**8**(5)

[65] Riaz S, Alam SS, Akhtar MW. Proteomic identification of human serum biomarkers in diabetes mellitus type 2. Journal of Pharmaceutical and Biomedical Analysis. 2010;**51**(5): 1103-1107. DOI: 10.1016/j.jpba.2009.11.016

[66] Rittner, Don, & Bailey, Ronald Albert. (2005). Encyclopedia of Chemistry (Science Encyclopedia).

[67] Robinson K. Homocysteine and Vascular Disease. 1st ed. Springer Science & Business Media; 2013

[68] Sacks FM, Hermans MP, Fioretto P, Valensi P, Davis T, Horton E, et al. Association between plasma triglycerides and high-density lipoprotein cholesterol and microvascular kidney disease and retinopathy in type 2 diabetes mellitus a global case–control study in 13 countries. Circulation. 2014;**129**(9): 999-1008

[69] Seemungal T, Wedzicha JA, Rios M. Homocysteine is Elevated in COPD. INTECH Open Access Publisher; 2012

[70] Shapiro AL, Viñuela E, Maizel JV. Molecular weight estimation of polypeptide chains by electrophoresis in SDS-polyacrylamide gels. Biochemical and Biophysical Research Communications. 1967;**28**(5):815-820

[71] Simon SA, Varadhan S. Ratio of triglycerides to high-density lipoprotein cholesterol and markers of liver injury in diabetes mellitus. Asian Journal of Pharmaceutical and Clinical Research. 2015;**8**(6)

[72] Sino, Biological Incorporated (Producer). Human folate receptor alpha protein (Folr1). 2016. [Image] Retrieved from: http://www.sinobiological.com/FOLR1-Protein-g-4464.html

[73] Strand TA, Taneja S, Kumar T, Manger MS, Refsum H, Yajnik CS. Vitamin B-12, folic acid, and growth in 6-to 30-month-old children: A randomized controlled trial. Pediatrics. 2015;**135**:e918-e926

[74] Vega GL, Barlow CE, Grundy SM, Leonard D, DeFina LF. Triglyceride–to–high-density-lipoprotein-cholesterol ratio is an index of heart disease mortality and of incidence of type 2 diabetes mellitus in men. Journal of Investigative Medicine. 2014;**62**(2): 345-349

[75] Wefuan, Fatimatou Kuhmbou. The Escalating Diabetes Epidemic: Determinants of Prevalence Disparity between Country Income Groups; 2013

[76] World Health Organization (WHO). Human vitamin and mineral requirements: FAO, Corporate Document Repository; 2001

[77] Wikipedia. Folic acid. 2016. Retrieved from: https://en.wikipedia.org/wiki/Folic_acid [26 May 2016]

[78] World Health Organization. Definition and Diagnosis of Diabetes Mellitus and Intermediate Hyperglycaemia: Report of a Wh. 2006

[79] Yi P, Melnyk S, Pogribna M, Pogribny IP, Hine RJ, James SJ. Increase in plasma homocysteine associated with parallel increases in plasma S-adenosylhomocysteine and lymphocyte DNA hypomethylation. Journal of Biological Chemistry. 2000;**275**(38): 29318-29323. DOI: 10.1074/jbc.M002725200

[80] Zheng M, Zhang M, Yang J, Zhao S, Qin S, Chen H, et al. Relationship between blood levels of methyl donor and folate and mild cognitive impairment in Chinese patients with type 2 diabetes: A case-control study. Journal of Clinical Biochemistry and Nutrition. 2014;**54**(2):122-128

Chapter 6

ZIKV Diagnostics: Current Scenario and Future Directions

Zilton Vasconcelos, Renata Campos Azevedo, Andrea Zin, Luiza Neves and Daniela Prado Cunha

Abstract

Since the discovery of the Zika virus in Uganda in 1947, diagnostic challenges remain, especially when we take into account the epidemiological context of the surveyed population. Genetic similarities with other flavivirus are responsible for cross-reactivity during serological evaluation that would often be the only resources to confirm the infection in asymptomatic cases or samples collected after the short window of viral RNA detection. The importance of Zika virus infection diagnosis is undoubtedly useful for pregnant women. This statement became evident after 2015 Brazil's Zika outbreak when a significant increase in cases of newborn with microcephaly was observed. Reverse transcriptase real-time PCR is the most reliable tool for Zika virus infection diagnosis. It detects viral RNA in both biological fluids and tissues and contributes to clinical case classification for initial description of developmental changes observed in neonates exposed congenitally to Zika virus. In conclusion, advances in serological diagnostic are urgent. The safest pathway for these studies requires laborious, subjective, and low throughput PRNT evaluations. Consequently, critical public health questions remain unanswered: how serum prevalent is the general population and pregnant women; can we define risk for congenital Zika syndrome (CZS) and Guillain-Barré syndrome; and how to assess vaccine efficacy and long-term protection.

Keywords: congenital Zika syndrome, real-time RT-PCR, PRNT, serum prevalence

1. Introduction

The Zika virus (ZIKV) belongs to the family *Flaviviridae*, genus *Flavivirus*, phylogenetically related to the dengue, yellow fever, and West Nile virus [1]. The main vector is the *Aedes aegypti*, hematophagous arthropod well adapted to urban living conditions, and for this reason, they are considered as emergent arboviruses, transmitted mainly by mosquitoes of the genus *Aedes.*

Clinically, the infection produces a self-limited febrile disease, whose symptoms of the acute phase are fever, headache, myalgia, and cutaneous rash. The virus incubation period in humans can range from 3 to 12 days, very similar to other arboviruses such as dengue and chikungunya virus. Thus, in regions where there is co-infection of these viruses, ZIKV can be sub-notified [2].

The ZIKV is transmitted to humans by the bite of infected female mosquitoes. However, other mechanisms of transmission have been described such as sexual transmission, blood transfusion, and through breastfeeding [3–6]. In addition, the presence of ZIKV RNA has been identified already in body fluids such as endocervical or nasopharyngeal swab, saliva, and urine, which highlights other possible routes of transmission [7–11]. For example, ZIKV RNA detection in amniotic fluid was essential to define fetal infection during vertical transmission [12].

During 1947, ZIKV was originally isolated from the serum of a febrile rhesus monkey female, which was used as a sentinel to detect wild yellow fever virus in the Zika forest, localized at Uganda [13]. At the same site, the ZIKV was also isolated from a mosquito genus *Aedes Africanus* sample. More than one decade later, the first case of ZIKV infection in humans was reported in Uganda, with a detailed clinical description performed by a doctor who described his own infection. He described symptoms such as a mild headache, diffuse maculopapular rash, fever, and malaise [14].

Despite this report, ZIKA infections remained relatively unknown until 2007, when it occurred as a major outbreak in Micronesia, representing the first evidence of virus circulation outside the African continent [15].

The disease spread rapidly across the islands of the Pacific Ocean, and in October 2013, initial cases were diagnosed in French Polynesia that culminates with a major epidemic [16]. In February 2014, for the first time, ZIKV was detected in the Americas, and those cases were reported on Easter Island (Chilean territory in the Pacific Ocean), probably related to the outbreak in Micronesia and French Polynesia [17].

In Brazil, the first autochthonous transmission was observed in April 2015 in patients from Rio Grande do Norte, where the sequence of a fragment of viral envelope protein revealed high identity with ZIKV sequences of the Asian lineage [2]. The Ministry of Health of Brazil confirmed the circulation of ZIKV in the country, after identifying the viral RNA in 16 serum samples (8 cases from Bahia and the others from Rio Grande do Norte). The main symptoms described were arthralgia, edema of the extremities, low fever, headache, and retro-orbital pain, conjunctival hyperemia, and maculopapular rashes, besides pruritus, dizziness, myalgia, and digestive disorder [18]. Approximately 80% of ZIKV infections are asymptomatic or display mild-to-moderate symptoms that are self-limited and last between 5 and 7 days [19].

However, before ZIKV outbreak detection, the number of infants born with microcephaly in the second half of 2015 increased tremendously, when compared with historical reports. This observation was done in the Northeast region of the country, and Recife and Salvador and their surroundings were the most severely affected cities [18]. Retrospectively, the anamnesis of most parturients revealed episodes of febrile illness and exanthematous skin lesions during pregnancy [20]. Considering the epidemiological situation, in November 2015, the World Health Organization confirmed the relationship between ZIKV infection and congenital malformations in neonates, with most cases developing microcephaly and/or ventriculomegaly [21, 22].

In addition to CNS malformations due to congenital infections, ZIKV outbreak was also associated with increased number of cases of Guillain-Barré Syndrome (GBS) in French Polynesia, indicating another CNS-related complication resulting from ZIKV infection [23, 24]. In Brazil, severe manifestations associated with ZIKV infections were also described such as GBS, transverse myelitis, and meningitis in adults [23, 25].

2. Structural properties, genetic context, and evolution

ZIKV particle is composed of a positive-sense, single-stranded RNA about 10,794 bp in length involved by a protein capsid and surrounding a lipid membrane. The RNA genome presents 5′ and 3′ untranslated regions (UTR), flanking a single long open reading frame (ORF) that translates one polyprotein (5'-C-prM-E-NS1-NS2A-NS2BNS3-NS4A-NS4B-NS5–3′). After cleavage, the polyprotein originates three structural proteins: capsid (C), membrane (M), and envelope (E), which are virus particle components, and seven nonstructural proteins: NS1, NS2A, NS2B, NS3, NS4A, NS4B, and NS5, which are involved in genome replication, polyprotein cleavage, and evasion of immunologic response (**Figure 1**) [26, 27]. The genome region, encoding flavivirus structural proteins, is more susceptible to genome variation when compared to nonstructural coding regions. Envelope protein (E), the major virion surface protein, is responsible for virus-cell interaction and membrane fusion, therefore being the major target for neutralizing antibodies [28]. Three domains were identified in E protein EDI, EDII, and EDIII. Karin Stettler and colleagues demonstrated that monoclonal antibodies (mAbs) directed to EDIII domain presented a potent and specific neutralizing activity to ZIKV, whereas mAbs directed to EDI/EDII domains partially neutralized ZIKV. Additionally, EDI/EDII-specific mAbs displayed higher cross-reactivity with dengue virus and enhanced *in vitro* viral infection, the phenomena described as antibody-dependent enhancement (ADE) [29]. This information should be considered in the development of ZIKV-specific diagnostic tool using antibody-based detection.

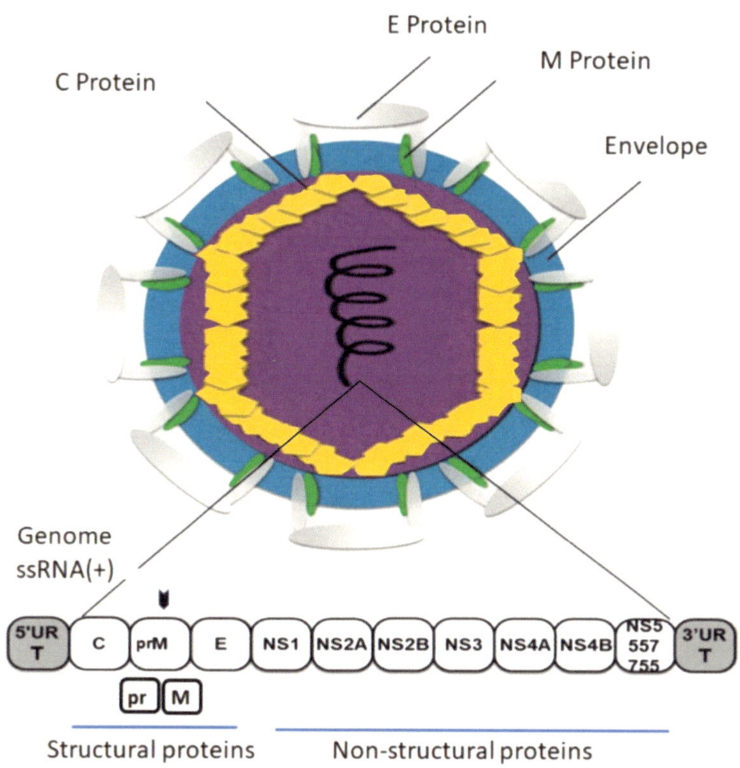

Figure 1.
Flavivirus particle diagram. Flavivirus are small spherical enveloped particles. Envelope protein (E), responsible for virus-cell interaction and membrane fusion, is the main target for neutralizing antibodies. The membrane-associated protein (M) results from the cleavage of the precursor to membrane protein (PrM) during virus maturation. Capsid proteins (C) assemble into an icosahedral structure (capsid) and involve the virus genomic material (ssRNA+). The genomic RNA coding region is flanked by two untranslated regions (5' UTR and 3' UTR) and produces a polyprotein that after proteolytic cleavage generates all functional virus proteins. Adapted from de Vasconcelos et al. [32].

Regarding the nonstructural proteins, NS1 glycoprotein is a multifunctional virulence factor that intracellularly co-localizes with the virus replication complex. In addition, an infected cell secretes NS1 as a hexameric lipoprotein, and several immunomodulatory functions are proposed for this particular protein [27]. The secretion of NS1 during acute dengue infection culminates in high serum levels of this protein. It occurs mainly during viremic period and encourages the development of NS1-based serologic diagnostic tests during previous dengue outbreaks [30]. Recently, one serologic test for ZIKV based on NS1 reinforces the use of this protein as an acute phase biomarker for flavivirus infection [31]. The NS5 protein is the largest viral protein whose C-terminal portion has RNA-dependent RNA polymerase (RdRP) activity. The polymerase region of the protein is generally conserved and can be used for the design of a broad range of primers.

Genetic studies using nucleotide sequences derived from the NS5 gene indicated three ZIKV lineages: East African, West African, and Asian [1]. However, after genome-based phylogenetic analysis, Asian and African strains remained as the only two genetic lineages for ZIKV [1]. ZIKV was associated with sporadic human infections in Africa and Southeast Asia by the end of the last century and was described as a zoonotic pathogen [33]. Due to the first isolation in monkeys and mosquitoes, the maintenance of the enzootic ZIKV was first attributed to those two groups of animals. However, antibodies have been detected in other animal species [26]. The ZIKV ability to switch species can play a special role in virus evolution. Especially in virus presenting RNA genome, nucleotide polymorphisms have long been noted as an evolutionary mechanism and indicate viral adaptation to the host [34, 35]. The mutation rate of ZIKV was demonstrated to be between 12 and 25 bases per year and corresponds to 0.2% of total ZIKV genome [36]. On top of the mutation rate, recombination events also play an important role in virus evolution. Fay et al. studying the molecular evolution of ZIKV reported a possible event of recombination in E and NS5 genes affecting African samples [28]. Han et al. propose a recombinant event within Brazilian strain (ZikaSPH2015), French Polynesia strain (H/PF/2013), and strain Z1106033 from Suriname [37]. Noteworthy, the recombination within Asian strains was not supported by Jun and colleagues when 196 ZIKV genomes were analyzed [38]. Therefore, further investigation is needed in order to accurately correlate genome changes and virulence.

3. Molecular tests (virus detection)

3.1 Quantitative polymerase chain reaction (qPCR)

Quantitative polymerase chain reaction amplification (qPCR) assays have been used for laboratory diagnosis of viral infectious diseases, since the sensitivity, specificity, and speed of the method allow the rapid detection of viruses in the host, human clinical samples, biopsies, and autopsy tissue. Since ZIKV is an RNA virus, running an initial reverse transcriptase step followed by a qPCR (RT-qPCR) is recommended. This method has been proved to be the most reliable one to confirm ZIKV infection during acute phase. Then, the first step for molecular ZIKV detection is the RNA isolation from different patient biological samples. When one is dealing with tissue samples, it is necessary to mechanically dissociate the freshly isolated tissue fragment to further conduct the nucleic acid extraction process. Other body fluid samples could be directly processed by an RNA isolation kit before viral genome reverse transcriptase reaction followed by amplification using ZIKV-specific primers and probes.

One of the limitations of the PCR technique is related to the presence of inhibitory substances, such as IgG, hemoglobin, lactoferrin in blood samples, and anticoagulants as heparin, that have already been described as PCR inhibitors [39]. In urine samples, the most critical component described as PCR inhibitors is urea, which can lead to degradation of the polymerase. However, its direct effect is dependent on urea concentration in the sample [40]. One way to monitor an inhibitory effect performed by PCR inhibitors is to carry out PCR positive control reactions. Spike-in experiments where an exogenous microorganism is added to your patient sample, to be co-extracted and amplified by another pair of primers and probe, could demonstrate inhibitors' presence when no amplification occurs [41].

Other limitations of the technique are related to both the time of infection (acute phase) and the type of sample chosen. This relationship is directly associated with the success of the laboratory confirmation. It is known that the presence of the virus in the plasma, saliva, breast milk, and nasopharyngeal swab is short, which allows the detection of viral RNA for a limited time, and during the acute phase of the disease, it is calculated on average 3–7 days after the occurrence of the symptoms. On the other hand, urine and saliva biological samples are optimal in the context of infant and newborn diagnosis due to low invasiveness and easy sample access [7, 10, 42–44]. In term of RNA detection, urine samples are described to present positivity from the fourth day until 29 days after the onset of symptoms [45], another time this sample opens the possibility of virus detection in longer times, even after the period of viremia. In adults, the biological sample that displays an increased period of viral persistence was semen, with cases where RNA detection occurred months after ZIKV infection. This raises the possibility that viral persistence could exist in both female and male genital tract (**Figure 2**) [8, 42, 46–48].

Apart of a positive RT-qPCR result for ZIKV, quantitative PCR offers the opportunity to deliver viral copies relative to tissue milligrams or fluid milliliters. To calculate viral load is imperative to construct a standard curve, from a serial dilution from a

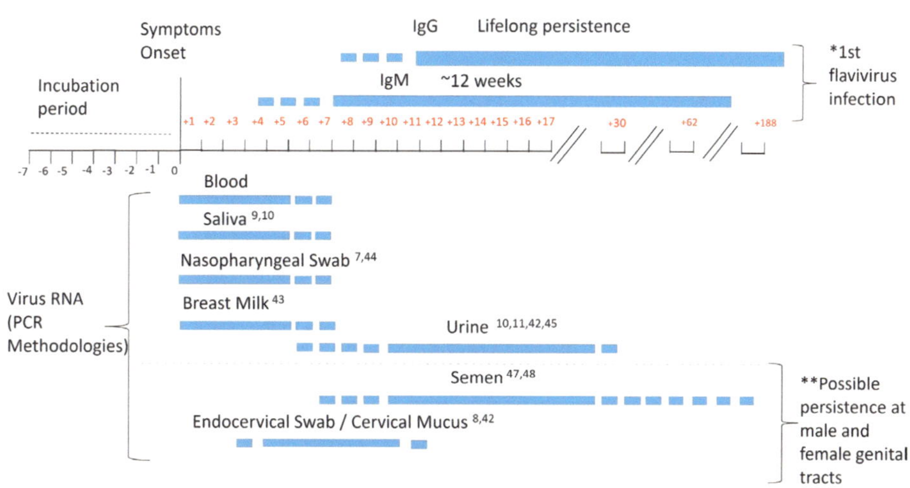

Figure 2.
*Literature evidence on the average period of virus RNA positivity in different body fluids and serum immunoglobulins levels after symptoms onset. Bars following the timeline indicate sample and proper molecular test that should be used to investigate infection after viral exposition, and dashes represent virus load decrease. *Immunoglobulins development expected for a primary flavivirus infected patient. A secondary exposition should promote earlier and higher IgG serum titles with short and reduced IgM levels. **ZIKV detection in genital fluids without viremia suggests viral tropism and persistence in those organs.*

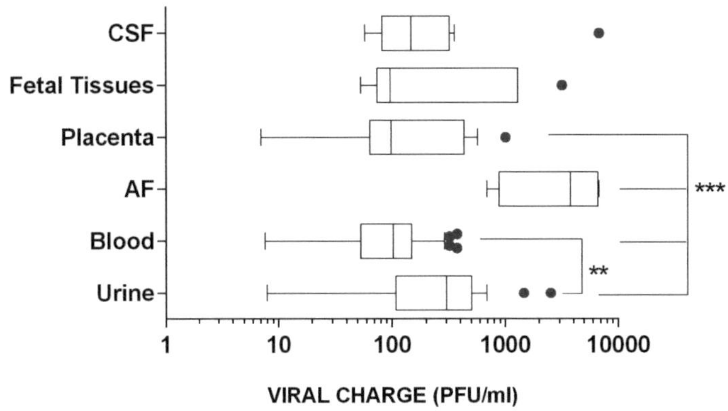

Figure 3.
*ZIKV RNA viral load in different biological samples. Noninvasive biological samples urine presents a higher viral load than blood samples (**p-value < 0.001). AF viral load is the highest when compared with other maternal samples (***p-value < 0.0001). AF, amniotic fluid. CSF, cerebrospinal fluid. Data generated from 13 placentas, 12 CSF, 6 fetal tissues, 4 AF, 24 urine, and 49 blood samples.*

reference sample that was previously determined RNA copies or plaque forming units (PFU). Then, the sample result will be possible to extrapolate from a relative measurement to an absolute quantification applied to a patient sample. This quantitative molecular diagnosis, in some pathologies (e.g., Hepatitis B and C, HIV, and herpes), has been already described as important indicators of health status and guide treatment choices. In the case of ZIKV infection, viral load does not appear to be related with clinical adverse events in congenital infections [49] or general population [50], but it was important to demonstrate viral load in different biological samples. As shown in **Figure 3**, the amniotic fluid has the highest viral load when compared with other biological fluids (urine or blood). However, in the context of an epidemiological outbreak, blood and urine are the more accessible samples with a significantly higher amount of virus and prolonged RNA detection (**Figure 2**) in urine than blood. The placenta did not present a significant difference when compared to blood and urine.

However, as discussed before, potential ZIKV RNA viral mutation turns RT-qPCR methods not infallible. Viral genetic evolution raises the possibility that polymorphisms inside primer annealing site may generate false negative amplification reactions of the ZIKV genome targeted region. During 2015 ZIKV outbreak, there was six widely used *in-house* RT-qPCR methods for diagnostic [1, 51–53], and their suitability for detection of ZIKV Asian strain was reported by Corman et al. with clear differences in sensitivity between them [54]. Poor PCR sensitivity and low RNA viral load observed in patient samples can be one of the explanations for the reduced number of laboratory-confirmed cases during the last epidemics. Few RT-qPCR kits for ZIKV detection were commercially available, and high costs also limited their use and could be the other reason for lower confirmed cases, especially in most of the endemic countries [55]. Besides the small amount of proficient clinical laboratories, performance control of *in-house* methodologies continues to be a challenging task; therefore, constant interlaboratory control should be stimulated, and standardized reagents, suitable probes, and primers should be used whenever possible.

3.2 Immunoenzymatic assays for virus detection

3.2.1 NS1 tests

Another flavivirus laboratory test that has gained prominence is the detection of NS1 viral protein in the serum, which in turn reflects a viral infection during the acute phase [56]. Largely used during DENV outbreaks, it is based on NS1 protein secretion by infected cells that reach high serum concentrations during periods of viral replication during flaviviruses and is considered an important acute phase biomarker for dengue fever. The assay format is suitable for use in routine laboratories around the world, allowing high-throughput testing in epidemic regions [31, 56]. Serological identification of ZIKV infections could also be maximized by parallel testing for IgM/IgG and NS1. The limitation of this assay is related to cross-reactivity between Zika and other flaviviruses, once they display a high degree of homology around their viral proteins, especially with dengue [1]. However, serological tests are constantly improving in order to increase the specificity through the use NS1-based anti-ZIKV, and it has been recently shown to be more sensitive and highly specific for the serodiagnosis of ZIKV infections but has not yet been commercially available [31].

3.2.2 Immunohistochemistry

Immunohistochemistry is an important tool to evaluate vertical transmission (placenta infection). Through this, viral particles of ZIKV have already been detected in placental defense cells, known as Hofbauer cells. These cells are of fetal origin and have the ability to migrate, in areas that have contact with trophoblastic epithelium and vessels of the fetus. It is believed that ZIKV has an immune system evasion mechanism and Hofbauer cells are responsible for viral dissemination, facilitating their access to the fetal compartment besides being a viral reservoir [57]. Due to access to the fetal compartment, ZIKV gains open access to fetal layer where no more immunological barrier could arrest their access to fetal central nervous system (CNS) [58].

ZIKV neuronal tropism was observed and described in affected fetus through necropsy studies. It was clearly demonstrated ZIKV antigens being expressed in large amounts in the brain with sporadic amounts of ZIKV antigens found in other fetal tissues, like liver, kidney, heart, and lung [59]. The opportunity to demonstrate CNS prolonged viral persistence was showed in a necropsy study conducted in a 5-month infant deceased after ZIKV congenital infection, which occurred in the first trimester of pregnancy. In this case report, ZIKV was still present in the brain tissue with no evidence of inflammation or brain damage evolution [60].

3.3 Virus isolation

Virus isolation using cell culture or animals remains as an option for virus detection. Vertebrate or invertebrate cells (e.g., *Aedes albopictus* clone C6/36) can be used for ZIKV propagation from the biological sample. This methodology allows the identification of infectious particles and was used to evaluate the body fluid potential to transmit the virus, such as saliva, breast milk, and urine. Additionally, cellular propagation increases the sensitivity in low viral load samples; however, the condition of sample transport and storage can impact the test result. This seems to be even more important for ZIKV, which demonstrates reduced stability in fluids and tissues [61]. Due to the time required for virus propagation and the cell culture structure need, viral isolation is not used in the diagnostic routine.

4. Serology tests

4.1 ELISA tests

The considerable cross-reactivity between flavivirus antibodies is the major challenge for serological test development. In endemic areas, where dengue and

Zika viruses cocirculate besides several proposed ELISA tests, individuals previously exposed to other flaviviruses are not prevented to be misclassified. Another complication for patients experiencing a second flavivirus infection is a shorten IgM antibody response that could misdiagnose the time for acute infection [1].

This presents a diagnostic dilemma for patients living in regions in which flaviviruses are endemic, and reliable diagnostic tools that are able to discriminate between primary and secondary ZIKV or DENV infections are urgently needed [1, 62, 63].

Although the precise period of onset and duration IgM class antibodies in response to ZIKV is not well defined, data known by other flaviviruses suggest that IgM for ZIKV should start to be detectable 7 days following infection, in the majority of symptomatic patients (**Figure 2**). However, it has been recently reported in the literature that seroconversion may occur earlier for one-third of patients [1, 42]. While anti-ZIKV antibody titers decrease, IgM to ZIKV has a window of detection more than 12 weeks following infection for over 80% of individuals [42]. According to the modeling studies with other flaviviruses, including West Nile virus (WNV) and dengue (DENV), this serum persistence is consistent with the antibody responses following infection, suggesting that the mean time to IgM seronegativity ranges from 5 to 6 months [64, 65].

Apart from initial antibody response, IgG neutralizing antibodies (NA) normally develop after IgM response and should persist for years to decades following primary infection. It is believed that NA to ZIKV directed to a key epitope after infection should be highly specific in patients without prior exposure to flaviviruses; however, NA specificity, in the setting of past exposure to a closely related flavivirus, such as DENV, once shares diverse antigenic determinants, decrease specificity to ZIKV epitopes [1].

Another tool that is being proposed in the context of outbreaks is rapid point of care (POC) tests which are based on the IgM/IgG immunochromatography assay. They should be used carefully and are normally interpreted as qualitative screening tests, and further serological investigations are needed. When they present a positive result, the patient samples should be sent for plaque reduction neutralization test (PRNT), or in the case of pregnant women, RT-qPCR is recommended [66, 67].

Special attention should be given to the risk group composed of newborns, where the immunoenzymatic serological test is extremely important to define congenital infection. ZIKV IgG antibody detection in a newborn may reflect maternal infection due to the passive transfer of this class of immunoglobulins. Based on other congenital infections, uninfected newborns display IgG levels that decline gradually in the first months of life, and this corresponds to the time of elimination of the antibody transmitted by the mother [68]. For this reason, tests to detect IgM are commonly used to confirm infection in the child, and in the context of flavivirus infection, only ZIKV was demonstrated to cross the placental and fetal immunological barriers [69, 70]. ZIKV IgM positivity in CSF, after congenital infection, was described as a reliable tool to discriminate fetal infection, and it characterized high risk for neuroinvasive disease [71].

4.2 Plaque reduction neutralization test (PRNT)

Until now, the unique reference standard for NA detection related to flaviviruses remains the plaque reduction neutralization test (PRNT). Although technically cumbersome, with several challenges: (1) long turnaround time of days to weeks; (2) requirement for live viral cultures; (3) technical-experience dependence; and (4) high degree of subjectivity, it offers the highest achievable level of specificity. This test initiates with a serial dilution of patient serum that is preincubated with live ZIKV or other closely related viruses (e.g., DENV), followed by deposition onto a virus-susceptible cell monolayer. After proper incubation time, any resulting plaques suggest nonneutralized live virus and absence of specific antibodies. Those plaques are quantified and compared to plaques number obtained in culture wells containing virus-only in order to establish which serum dilution promotes 90% reduction in plaques occurrence in this patient sample (PRNT90). ZIKV and DENV PRNT90 titers are subsequently compared to define the specificity of the NAs [72]. Although the PRNT assay has a higher sensitivity than other tests such as hemagglutination and enzyme immunoassay and is more specific than other serological methods for the diagnosis, the limitation of this test is not discriminate between antibody classes and, especially in secondary flavivirus infections, cross-reactive neutralizing antibodies may contribute to virus neutralization [73]. In addition, PRNTs are time-consuming and laborious and are to be restricted to reference laboratories and research centers because they require the technical capacity to carry them out, as well as a complex laboratory structure [46].

5. Clinical context

Zika virus (ZIKV) infection is usually an asymptomatic or a mild symptomatic disease in adults, with maculopapular and pruritic rash, fever, conjunctivitis, joint pain, headache, and muscle pain [74]. However, infection during pregnancy may be transmitted to the fetus and causes severe systemic fetuses' malformations, comprising the congenital Zika syndrome (CZS).

The clinical features of CZS have been described since 2015 Zika's outbreak in Brazil. They can be divided into structural and functional components [75]. The structural components are cranial morphology, brain, and ocular anomalies, as well as congenital contractures [75]. The functional component is related to neurologic sequelae [75]. Therefore, CZS may consist of (A) cranial morphology: severe microcephaly, overlapping cranial sutures, prominent occipital bone, redundant scalp skin, and neurologic impairment; (B) brain anomalies: thin cerebral cortices, abnormal gyral patterns, increased fluid spaces, subcortical calcifications, corpus callosum abnormalities, decreased white matter, and cerebellar hypoplasia; (C) ocular anomalies: macular scarring (**Figure 4A**), focal pigmentary retinal mottling (**Figure 4B**), and optic nerve hypoplasia or atrophy (**Figure 4A**); (D) congenital contractures: arthrogryposis and club feet; and (E) neurologic sequelae: marked early hypertonia, symptoms of extrapyramidal involvement, epilepsy, and irritability [75–77]. Early recognition and referral to multidisciplinary care may result in a better outcome for each one of the abnormalities described.

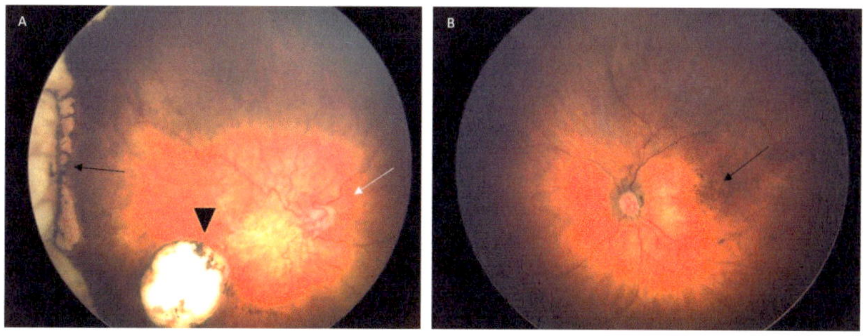

Figure 4.
(A) Right eye retinography showing macular (black arrowhead) and peripheral (black arrow) scarring and optic nerve hypoplasia (white arrow); (B) left eye retinography showing focal pigmentary retinal mottling (black arrow).

The diagnostic approach during prenatal care should be different in an endemic area and in a nonendemic area. Endemic area means residence in or travels to the affected area. In addition, in each of these two scenarios, the diagnosis is divided into symptomatic and asymptomatic pregnant women.

5.1 Endemic area (residence or travel)

5.1.1 Symptomatic pregnant women

The ZIKV infection is divided into acute and convalescent phases. The acute phase is within the first 7 days of the symptoms, and the convalescent phase is 2–12 weeks after [78]. As shown in **Figure 5**, during the acute phase, the possible primary tests are the ZIKV RT-qPCR and the Immunoglobulin (Ig) M. The first one can be obtained from serum, saliva, urine, or amniotic fluid. When it is detected in the serum or saliva, it is confined to 5–7 days after onset of symptoms, and when it is detected in urine, it lasts up to 3 weeks after onset of illness. The evidence is lacking on the diagnostic accuracy of RT-qPCR of amniotic fluid and on the optimal time to perform amniocentesis [74]. The IgM is detectable ≥4 days after onset of illness [78]. In the convalescent phase, the primary test is IgM and the second test is plaque reduction neutralization test (PRNT). However, the Zika virus RT-qPCR can also be performed during this phase.

If the primary results are **negative ZIKV RT-qPCR and negative ZIKV IgM** (red flowchart in **Figure 5**), there is no evidence of virus detection, but an intrauterine ultrasound scan (US) should be done. If there are any fetal development abnormalities, serological tests for others congenital infections could be done and the pregnant woman should be referred to a specialized care. The possible congenital infections are toxoplasmosis, rubella, cytomegalovirus (CMV), herpes virus, parvovirus B19, varicella, and syphilis, known by the acronym TORCH's syndrome (Sd).

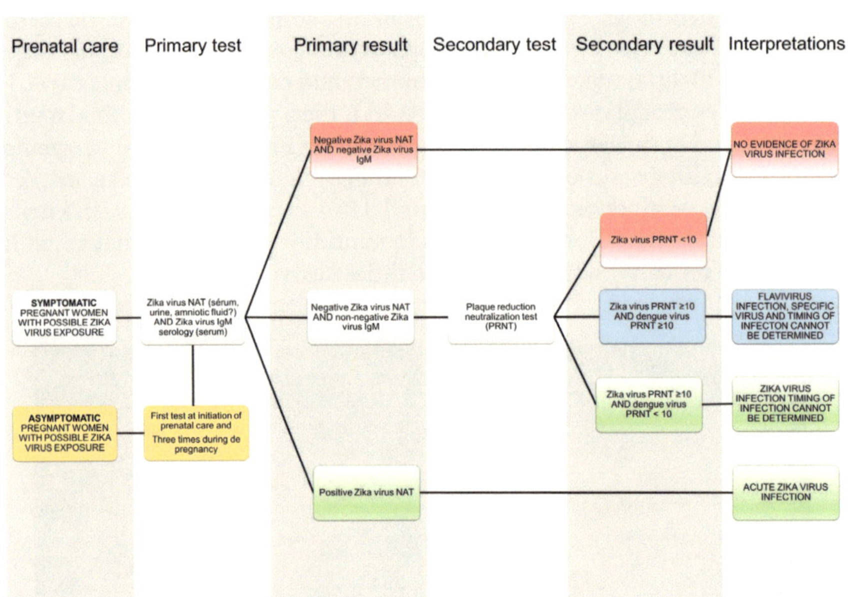

Figure 5.
Flowchart showing initially two clinical scenarios (symptomatic and asymptomatic pregnant women) and the different primary and secondary tests and results for Zika virus diagnosis during pregnancy. NAT: RT-qPCR nucleic acid test. IgM: Immunoglobulin M. US: Intrauterine ultrasound. PRNT: Plaques reduction neutralization test.

The possible fetal abnormalities are microcephaly, intracranial calcifications, ocular lesions and calcifications, ventriculomegaly, abnormal sulcation and gyration, abnormal cortical development (lissencephaly), cerebral atrophy, callosal dysgenesis, cerebellar atrophy, brainstem hypoplasia, microphthalmia and arthrogryposis, intrauterine growth restriction, and evidence of placental insufficiency [74]. If there is absence of fetal abnormalities, the pregnant can follow routine antenatal care and it should be considered to repeat the US at 28–30 weeks [74].

If the primary results are **negative ZIKV RT-qPCR and non-negative ZIKV IgM**, the secondary test PRNT should be performed. It can be done for any other flavivirus that might be found in that geographic area [74]. If ZIKV PRNT ≤10 (white flowchart in **Figure 5**), there is no evidence of Zika virus infection. If ZIKV PRNT ≥10 and dengue virus PRNT ≥10 (blue flowchart in **Figure 5**), there is evidence of a flavivirus infection but no confirmation of a specific virus. Finally, if ZIKV PRNT ≥10 and dengue virus PRNT <10 (green flowchart in **Figure 5**), there is evidence of ZIKV infection. In any of the last two scenarios, the timing of infection cannot be determined.

If the primary result is **positive ZIKV RT-qPCR** (green flowchart in **Figure 5**), acute ZIKV infection is confirmed and US should be done. If there is any fetal abnormality, the pregnant woman should be referred to a specialized care, and if there is no abnormality, routine antenatal care should be done, as well as a US follow-up every 4 weeks until birth [74, 79].

5.1.2 Asymptomatic pregnant women

In endemic area of residence or traveling, the pregnant woman with possible ZIKV exposure should be first tested at initiation of prenatal care and three times during pregnancy [67]. The primary tests should be ZIKV RT-qPCR and ZIKV IgM serology. If available, US should be considered for fetal morphology assessment. The possible results of the laboratory tests and US are described above.

5.2 Nonendemic area

The asymptomatic woman should follow the routine antenatal care. The symptomatic woman should do the primary tests ZIKV RT-qPCR and ZIKV IgM serology. If available, US for fetal morphology assessment should be considered.

Additionally, the differential diagnosis is also crucial for the correct assessment and management of the disease. Signs of maternal disease, as well as intrauterine and neonatal findings, are similar in most of the TORCH's congenital infections [75, 80]. Moreover, some genetic syndromes (Sd), such as Aicardi-Goutières Sd, pseudo-TORCH Sd, and mutations in the JAM3, NDE1, and ANKLE2 genes can also be in the differential diagnosis [75].

The main clinical findings in maternal TORCH infections include nonspecific signs, like fever, fatigue, malaise, headache, myalgia, arthralgia, lymphadenopathy, conjunctivitis, and making the ability to recognize the specific etiological agent difficult. Moreover, there are typical findings in some of these infections that can lead to the diagnosis. In herpes simplex virus infection, there is also painful genital ulcer, pruritus, and dysuria [80]. In primary syphilis infection, there is a firm, round, and painless chancre on external genitals and vagina, lasting 3–6 weeks [81]. In rubella, parvovirus B19, varicella, and ZIKV, cutaneous rash is an additional important sign [80, 82, 83]. The rash in rubella is maculopapular and usually progresses from the face to the body, fading within 2–4 days [80, 82, 83]. In parvovirus B19, the rash is also maculopapular, persisting until the 6th day and disappearing on the 7–9th day after the first appearance [80, 82, 83]. First, it occurs on the face, sparing nasal, and periorbital zones and second, on the trunk and extremities [83]. In varicella, it is initially pruritic and maculopapular, following a vesicular phase until crushing occurs, usually about 5 days later [84]. Finally, the rash in ZIKV infection is pruritic, descending, and maculopapular, beginning proximally and spreading to the extremities with resolution occurring within 1–4 days of onset [14, 74].

Nonspecific intrauterine US findings in TORCH infections are intrauterine growth restriction, abnormal fluid imbalance (ascites, hydrops, pericardial effusion, pleural effusion, and oligo-polyhydramnios), hepatosplenomegaly and hepatic calcifications, echogenic bowel, echogenic kidneys, and limb deformities [80, 85]. Cerebral brain calcification, although relatively common in most of these infections, can be an important distinguishing factor due to its location [80, 85]. In congenital toxoplasmosis, the calcifications are diffuse and widely distributed, and in rubella, they are at basal ganglia [80, 85]. In CMV congenital infection, the calcifications are usually punctate and periventricular or cortical [80, 85]. Conversely, in ZIKV, the calcifications are larger and denser, usually appearing at the gray and white matter junction [80, 86].

As well as seen with maternal clinical signs and intrauterine US findings, TORCH congenital infection can also produce a similar neonatal clinical presentation, although the features rarely occur simultaneously [87, 88]. The common clinical features include growth retardation, prematurity, jaundice, anemia, hepatosplenomegaly, microcephaly, cerebral calcifications, chorioretinitis, cataracts, microphthalmia, and others [87, 88]. However, clinical findings are more associated, although not exclusively, with a determined etiological agent. Cardiac lesions and deafness have been described in CMV and rubella congenital infections. Cicatricial skin lesions and limb hypoplasia are seen in varicella virus congenital infection [87]. Petechial or purpuric form exanthema (blueberry muffin spot) is typical of CMV infection [80, 82]. Chorioretinitis, hydrocephalus, and cerebral calcifications comprise the characteristic triad of congenital toxoplasmosis [89].

Finally, clinical signs of CZS can also be found in other TORCH infections. Cranial ZIKV morphology and brain anomalies can occur in congenital CMV infection [75]. Congenital contractures can appear in congenital rubella, varicella, and Coxsackie B infections [75]. Neurologic sequelae are seen in most TORCH infections [75]. Ocular anomalies, such as pigmentary mottling and chorioretinal scars, can manifest differently in each of these congenital infections. In rubella, the pigment mottling is usually diffused compared with the focal pigment mottling seen in Zika virus infection (**Figure 4B**) [77]. Chorioretinal lesions in toxoplasmosis can present with active exudative retinitis or regressed macular or peripapillary retinal scar [77]. In ZIKV congenital infection, chorioretinal lesions are atrophic and colobomatous-like and are found in the macula or retinal periphery (**Figure 4A**) [77]. Optic nerve hypoplasia, commonly seen in ZIKV congenital infection, is seldom seen in rubella, toxoplasmosis, herpes, and CMV congenital infections [77].

6. Conclusions

Human clinical diagnosis of ZIKV infection in regions where other arboviruses circulate, mainly DENV and CHIKV, has become a daunting task; therefore, laboratory confirmation is crucial for conclusive diagnosis. Detection of virus genome by RT-qPCR is helpful and demonstrated to be a reliable tool; however, the limited window for virus detection, low viral load, and stability restricted the use of these methodologies. Antibody screening is also hampered by cross-reactivity among other flaviviruses.

Due to impact of ZIKV infection, especially for pregnant women, a confident ZIKV serology test is urgently needed, and this will promote a better prenatal follow-up, especially in endemic areas. Nonetheless, as mentioned above, there are typical findings to aid the correct diagnosis and, thus, the appropriate management and counseling of the disease.

Acknowledgements

We thank Fernandes Figueira Institute Pediatric Infectious Disease Team for helpful discussions during this work and field expertise from ZIKVIRUSIFF Cohort study registered on clinicaltrials.gov with identifier NCT03255369.

This work was partially supported by Fernandes Figueira Institute Research Program (Programa de Incentivo à Pesquisa II), Rio de Janeiro State Research Grant (Fundação de Amparo à Pesquisa do Estado do Rio de Janeiro—grant E-26/201.840/2017), National Research Grant (Conselho Nacional de Desenvolvimento Científico e Tecnológico—grant 441098/2016-9), and the Coordenação de Aperfeiçoamento de Pessoal de Nível Superior—Brasil (CAPES)—Finance Code 001.

Conflict of interest

The authors declare no conflict of interest.

Author details

Zilton Vasconcelos[1]*, Renata Campos Azevedo[2], Andrea Zin[1], Luiza Neves[1] and Daniela Prado Cunha[1]

1 Fernandes Figueira Institute, Fiocruz, Rio de Janeiro, Brazil

2 Institute of Microbiology Paulo de Góes, Federal University of Rio de Janeiro, Rio de Janeiro, Brazil

*Address all correspondence to: zvasconcelos@iff.fiocruz.br

References

[1] Lanciotti RS, Kosoy OL, Laven JJ, Velez JO, Lambert AJ, Johnson AJ, et al. Genetic and serologic properties of Zika virus associated with an epidemic, Yap State, Micronesia, 2007. Emerging Infectious Diseases [Internet]. 2008;**14**(8):1232-1239. DOI: 10.3201/eid1408.080287

[2] Zanluca C, Melo VCA de, Mosimann ALP, Santos GIV dos, Santos CND dos, Luz K. First report of autochthonous transmission of Zika virus in Brazil. The Memórias do Instituto Oswaldo Cruz [Internet]. 2015;**110**(4):569-572. Available from: http://www.scielo.br/scielo.php?script=sci_abstract&pid=S0074-02762015000400569&lng=en&nrm=iso&tlng=en [cited: 27 August 2018]

[3] Atkinson B, Graham V, Miles RW, Lewandowski K, Dowall SD, Pullan ST, et al. Complete genome sequence of Zika virus isolated from semen. Genome Announcements [Internet]. 2016;**4**(5):1-2. DOI: 10.1128/genomeA.01116-16

[4] Musso D, Nhan T, Robin E, Roche C, Bierlaire D, Zisou K, et al. Potential for Zika virus transmission through blood transfusion demonstrated during an outbreak in French Polynesia, November 2013 to February 2014. Euro Surveillance [Internet]. 2014;**19**(14):1-3. Available from: https://www.ncbi.nlm.nih.gov/pubmed/24739982

[5] De Carvalho NS, De Carvalho BF, Fugaça CA, Dóris B, Biscaia ES. Zika virus infection during pregnancy and microcephaly occurrence: A review of literature and Brazilian data. The Brazilian Journal of Infectious Diseases [Internet]. 2016;**20**(3):282-289. DOI: 10.1016/j.bjid.2016.02.006

[6] Mann TZ, Haddad LB, Williams TR, Hills SL, Read JS, Dee DL, et al. Breast milk transmission of flaviviruses in the context of Zika virus: A systematic review. Paediatric and Perinatal Epidemiology [Internet]. 2018;**32**(4):358-368. DOI: 10.1111/ppe.12478

[7] Fonseca K, Drebot M, MacDonald J, Lindsay R, Pabbaraju K, Tellier R, et al. First case of Zika virus infection in a returning Canadian traveler. The American Journal of Tropical Medicine and Hygiene [Internet]. 2014;**91**(5):1035-1038. DOI: 10.4269/ajtmh.14-0151

[8] Visseaux B, Mortier E, Houhou-Fidouh N, Brichler S, Collin G, Larrouy L, et al. Zika virus in the female genital tract. Lancet Infectious Diseases [Internet]. 2016;**16**(11):1220. DOI: 10.1016/s1473-3099(16)30387-5

[9] Barzon L, Pacenti M, Berto A, Sinigaglia A, Franchin E, Lavezzo E, et al. Isolation of infectious Zika virus from saliva and prolonged viral RNA shedding in a traveller returning from the Dominican Republic to Italy, January 2016. Euro Surveillance [Internet]. 2016;**21**(10):1-5. DOI: 10.2807/1560-7917.es.2016.21.10.30159

[10] Bonaldo MC, Ribeiro IP, Lima NS, Dos Santos AAC, Menezes LSR, da Cruz SOD, et al. Isolation of infective Zika virus from urine and saliva of patients in Brazil. PLoS Neglected Tropical Diseases [Internet]. 2016;**10**(6):e0004816. DOI: 10.1371/journal.pntd.0004816

[11] Gourinat A-C, O'Connor O, Calvez E, Goarant C, Dupont-Rouzeyrol M. Detection of Zika virus in urine. Emerging Infectious Diseases [Internet]. 2015;**21**(1):84-86. DOI: 10.3201/eid2101.140894

[12] da Silva SR, Cheng F, Gao S-J. Zika Virus and Diseases: From Molecular Biology to Epidemiology [Internet]. New Jersey, United States: John

Wiley & Sons; 2018. 264 p. Available from: https://market.android.com/details?id=book-EWxODwAAQBAJ

[13] Dick GWA, Kitchen SF, Haddow AJ. Zika virus. I. Isolations and serological specificity. Transactions of the Royal Society of Tropical Medicine and Hygiene [Internet]. 1952;**46**(5):509-520. Available from: https://www.ncbi.nlm.nih.gov/pubmed/12995440

[14] Simpson DIH. Zika virus infection in man. Transactions of the Royal Society of Tropical Medicine and Hygiene [Internet]. 1964;**58**(4):339-348. DOI: 10.1016/0035-9203(64)90201-9

[15] Duffy MR, Chen T-H, Hancock WT, Powers AM, Kool JL, Lanciotti RS, et al. Zika virus outbreak on Yap Island, Federated States of Micronesia. New England Journal of Medicine [Internet]. 2009;**360**(24):2536-2543. DOI: 10.1056/NEJMoa0805715

[16] Cao-Lormeau V-M, Roche C, Teissier A, Robin E, Berry A-L, Mallet H-P, et al. Zika virus, French polynesia, South pacific, 2013. Emerging Infectious Diseases [Internet]. 2014;**20**(6):1085-1086. DOI: 10.3201/eid2006.140138

[17] Musso D, Nilles EJ, Cao-Lormeau V-M. Rapid spread of emerging Zika virus in the Pacific area. Clinical Microbiology and Infection [Internet]. 2014;**20**(10):O595-O596. DOI: 10.1111/1469-0691.12707

[18] Mlakar J, Korva M, Tul N, Popović M, Poljšak-Prijatelj M, Mraz J, et al. Zika virus associated with microcephaly. New England Journal of Medicine [Internet]. 2016;**374**(10):951-958. DOI: 10.1056/NEJMoa1600651

[19] Plourde AR, Bloch EM. A literature review of Zika virus. Emerging Infectious Diseases [Internet]. 2016;**22**(7):1185-1192. DOI: 10.3201/eid2207.151990

[20] Calvet G, Aguiar RS, Melo ASO, Sampaio SA, de Filippis I, Fabri A, et al. Detection and sequencing of Zika virus from amniotic fluid of fetuses with microcephaly in Brazil: A case study. Lancet Infectious Diseases [Internet]. 2016;**16**(6):653-660. DOI: 10.1016/s1473-3099(16)00095-5

[21] Štrafela P, Vizjak A, Mraz J, Mlakar J, Pižem J, Tul N, et al. Zika virus-associated micrencephaly: A thorough description of neuropathologic findings in the fetal central nervous system. Archives of Pathology & Laboratory Medicine [Internet]. 2017;**141**(1):73-81. DOI: 10.5858/arpa.2016-0341-SA

[22] Del Campo M, Feitosa IML, Ribeiro EM, Horovitz DDG, Pessoa ALS, França GVA, et al. Zika Embryopathy task force-Brazilian Society of Medical Genetics ZETF-SBGM. The phenotypic spectrum of congenital Zika syndrome. American Journal of Medical Genetics. Part A [Internet]. 2017;**173**(4):841-857. DOI: 10.1002/ajmg.a.38170

[23] Oehler E, Watrin L, Larre P, Leparc-Goffart I, Lastère S, Valour F, et al. Zika virus infection complicated by Guillain-Barré syndrome—Case report, French Polynesia, December 2013. Euro Surveillance [Internet]. 2014;**19**(9):1-3. DOI: 10.2807/1560-7917.es2014.19.9.20720

[24] Avelino-Silva VI, Martin JN. Association between Guillain-Barré syndrome and Zika virus infection. Lancet [Internet]. 2016;**387**(10038):2599. DOI: 10.1016/s0140-6736(16)30843-1

[25] Vasconcelos PF da C, da Costa Vasconcelos PF. Doença pelo vírus Zika: um novo problema emergente nas Américas? Revista Pan-Amazônica de Saúde [Internet]. 2015;**6**(2):9-10. DOI: 10.5123/s2176-62232015000200001

[26] Haddow AD, Schuh AJ, Yasuda CY, Kasper MR, Heang V, Huy R, et al.

Genetic characterization of Zika virus strains: Geographic expansion of the Asian lineage. PLoS Neglected Tropical Diseases [Internet]. 2012;**6**(2):e1477. DOI: 10.1371/journal.pntd.0001477

[27] Freire MCLC, Pol-Fachin L, Coêlho DF, Viana IFT, Magalhães T, Cordeiro MT, et al. Mapping putative B-cell Zika virus NS1 epitopes provides molecular basis for anti-NS1 antibody discrimination between Zika and dengue viruses. ACS Omega [Internet]. 2017;**2**(7):3913-3920. DOI: 10.1021/acsomega.7b00608

[28] Faye O, Freire CCM, Faye O, de Oliveira JV, Zanotto PMA, Diallo M, et al. Molecular evolution of Zika virus during its emergence in the 20th century. International Journal of Infectious Diseases [Internet]. 2014;**21**:2-3. DOI: 10.1016/j.ijid.2014.03.411

[29] Stettler K, Beltramello M, Espinosa DA, Graham V, Cassotta A, Bianchi S, et al. Specificity, cross-reactivity, and function of antibodies elicited by Zika virus infection. Science [Internet]. 2016;**353**(6301):823-826. DOI: 10.1126/science.aaf8505

[30] Zhang H, Li W, Wang J, Peng H, Che X, Chen X, et al. NS1-based tests with diagnostic utility for confirming dengue infection: A meta-analysis. International Journal of Infectious Diseases [Internet]. 2014;**26**:57-66. DOI: 10.1016/j.ijid.2014.02.002

[31] Balmaseda A, Stettler K, Medialdea-Carrera R, Collado D, Jin X, Zambrana JV, et al. Antibody-based assay discriminates Zika virus infection from other flaviviruses. Proceedings of the National Academy of Sciences of the United States of America [Internet]. 2017;**114**(31):8384-8389. DOI: 10.1073/pnas.1704984114

[32] de Vasconcelos ZFM, Azevedo RC, Thompson N, Gomes L, Guida L, Moreira MEL. Challenges for molecular and serological ZIKV infection confirmation. Child's Nervous System [Internet]. 2018;**34**(1):79-84. DOI: 10.1007/s00381-017-3641-5

[33] Hayes EB. Zika virus outside Africa. Emerging Infectious Diseases [Internet]. 2009;**15**(9):1347-1350. DOI: 10.3201/eid1509.090442

[34] Longdon B, Brockhurst MA, Russell CA, Welch JJ, Jiggins FM. The evolution and genetics of virus host shifts. PLoS Pathogens [Internet]. 2014;**10**(11):e1004395. DOI: 10.1371/journal.ppat.1004395

[35] Plotkin JB, Kudla G. Synonymous but not the same: The causes and consequences of codon bias. Nature Reviews. Genetics [Internet]. 2011;**12**(1):32-42. DOI: 10.1038/nrg2899

[36] Logan IS. ZIKA—How fast does this virus mutate? Dongwuxue Yanjiu [Internet]. 2016;**37**(2):110-115. DOI: 10.13918/j.issn.2095-8137.2016.2.110

[37] Han J-F, Jiang T, Ye Q, Li X-F, Liu Z-Y, Qin C-F. Homologous recombination of Zika viruses in the Americas. The Journal of Infection [Internet]. 2016;**73**(1):87-88. DOI: 10.1016/j.jinf.2016.04.011

[38] Jun S-R, Wassenaar TM, Wanchai V, Patumcharoenpol P, Nookaew I, Ussery DW. Suggested mechanisms for Zika virus causing microcephaly: What do the genomes tell us? BMC Bioinformatics [Internet]. 2017;**18**(S14):81-92. DOI: 10.1186/s12859-017-1894-3

[39] Al-Soud WA, Rådström P. Purification and characterization of PCR-inhibitory components in blood cells. Journal of Clinical Microbiology [Internet]. 2001;**39**(2):485-493. DOI: 10.1128/JCM.39.2.485-493.2001

[40] Wilson IG. Inhibition and facilitation of nucleic acid

amplification. Applied and Environmental Microbiology [Internet]. 1997;**63**(10):3741-3751. Available from: https://www.ncbi.nlm.nih.gov/pubmed/9327537

[41] Rådström P, Knutsson R, Wolffs P, Lövenklev M, Löfström C. Pre-PCR processing: Strategies to generate PCR-compatible samples. MBio Journal [Internet]. 2004;**26**(2):133-146. Available from: http://link.springer.com/10.1385/MB:26:2:133

[42] Paz-Bailey G, Rosenberg ES, Doyle K, Munoz-Jordan J, Santiago GA, Klein L, et al. Persistence of Zika virus in body fluids—Preliminary report. The New England Journal of Medicine [Internet]. 14 Feb 2017;**379**(13):1234-1243. DOI: 10.1056/NEJMoa1613108

[43] Colt S, Garcia-Casal MN, Peña-Rosas JP, Finkelstein JL, Rayco-Solon P, Weise Prinzo ZC, et al. Transmission of Zika virus through breast milk and other breastfeeding-related bodily-fluids: A systematic review [Internet]. In: Horstick O, editor. PLOS Neglected Tropical Diseases. 10 Apr 2017;**11**(4):e0005528. DOI: 10.1371/journal.pntd.0005528. eCollection 2017 Apr

[44] Leung GHY, Baird RW, Druce J, Anstey NM. ZIKA virus infection in Australia following a monkey bite in Indonesia. The Southeast Asian Journal of Tropical Medicine and Public Health [Internet]. 2015;**46**(3):460-464. Available from: https://www.ncbi.nlm.nih.gov/pubmed/26521519

[45] Campos R de M, Cirne-Santos C, Meira GLS, Santos LLR, de Meneses MD, Friedrich J, et al. Prolonged detection of Zika virus RNA in urine samples during the ongoing Zika virus epidemic in Brazil. Journal of Clinical Virology [Internet]. 2016;**77**: 69-70. DOI: 10.1016/j.jcv.2016.02.009

[46] Landry ML, St George K. Laboratory diagnosis of Zika virus infection. Archives of Pathology & Laboratory Medicine [Internet]. 2017;**141**(1):60-67. DOI: 10.5858/arpa.2016-0406-SA

[47] Nicastri E, Castilletti C, Liuzzi G, Iannetta M, Capobianchi MR, Ippolito G. Persistent detection of Zika virus RNA in semen for six months after symptom onset in a traveller returning from Haiti to Italy, February 2016. Euro Surveillance [Internet]. 2016;**21**(32):1-4. DOI: 10.2807/1560-7917.ES.2016.21.32.30314

[48] Atkinson B, Hearn P, Afrough B, Lumley S, Carter D, Aarons EJ, et al. Detection of Zika virus in semen. Emerging Infectious Diseases [Internet]. 2016;**22**(5):940. DOI: 10.3201/eid2205.160107

[49] Halai U-A, Nielsen-Saines K, Moreira ML, de Sequeira PC, Junior JPP, de Araujo Zin A, et al. Maternal Zika virus disease severity, virus load, prior dengue antibodies, and their relationship to birth outcomes. Clinical Infectious Diseases [Internet]. 2017;**65**(6):877-883. DOI: 10.1093/cid/cix472

[50] Musso D, Rouault E, Teissier A, Lanteri MC, Zisou K, Broult J, et al. Molecular detection of Zika virus in blood and RNA load determination during the French Polynesian outbreak. Journal of Medical Virology [Internet]. 2017;**89**(9):1505-1510. DOI: 10.1002/jmv.24735

[51] Faye O, Faye O, Diallo D, Diallo M, Weidmann M, Sall A. Quantitative real-time PCR detection of Zika virus and evaluation with field-caught mosquitoes. Virology Journal [Internet]. 2013;**10**(1):311. DOI: 10.1186/1743-422x-10-311

[52] Pyke AT, Daly MT, Cameron JN, Moore PR, Taylor CT, Hewitson GR, et al. Imported Zika virus infection from the Cook Islands into Australia, 2014. PLOS Currents [Internet]. 2014;**2**:6.

DOI: 10.1371/currents.outbreaks.4635a54dbffba2156fb2fd76dc49f65e

[53] Tappe D, Nachtigall S, Kapaun A, Schnitzler P, Günther S, Schmidt-Chanasit J. Acute Zika virus infection after travel to Malaysian Borneo, September 2014. Emerging Infectious Diseases [Internet]. 2015;**21**(5):911-913. DOI: 10.3201/eid2105.141960

[54] Corman VM, Rasche A, Baronti C, Aldabbagh S, Cadar D, Reusken CB, et al. Assay optimization for molecular detection of Zika virus. Bulletin of the World Health Organization [Internet]. 2016;**94**(12):880-892. DOI: 10.2471/BLT.16.175950

[55] Waggoner JJ, Pinsky BA. Zika virus: Diagnostics for an emerging pandemic threat. Journal of Clinical Microbiology [Internet]. 2016;**54**(4):860-867. DOI: 10.1128/JCM.00279-16

[56] Balmaseda A, Zambrana JV, Collado D, García N, Saborío S, Elizondo D, et al. Comparison of four serological methods and two reverse transcription-PCR assays for diagnosis and surveillance of Zika virus infection. Journal of Clinical Microbiology [Internet]. 2018;**56**(3):1-13. DOI: 10.1128/JCM.01785-17

[57] Rosenberg AZ, Yu W, Hill DA, Reyes CA, Schwartz DA. Placental pathology of Zika virus: Viral infection of the placenta induces villous stromal macrophage (Hofbauer cell) proliferation and hyperplasia. Archives of Pathology & Laboratory Medicine [Internet]. 2017;**141**(1):43-48. DOI: 10.5858/arpa.2016-0401-OA

[58] Bhatnagar J, Rabeneck DB, Martines RB, Reagan-Steiner S, Ermias Y, Estetter LBC, et al. Zika virus RNA replication and persistence in brain and placental tissue. Emerging Infectious Diseases [Internet]. 2017;**23**(3):405-414. DOI: 10.3201/eid2303.161499

[59] Azevedo RSS, Araujo MT, Martins Filho AJ, Oliveira CS, Nunes BTD, Cruz ACR, et al. Zika virus epidemic in Brazil. I. Fatal disease in adults: Clinical and laboratorial aspects. Journal of Clinical Virology [Internet]. 2016;**85**: 56-64. DOI: 10.1016/j.jcv.2016.10.024

[60] Chimelli L, Pone SM, Avvad-Portari E, Vasconcelos ZFM, Zin AA, Cunha DP, et al. Persistence of Zika virus after birth: Clinical, virological, neuroimaging, and neuropathological documentation in a 5-month infant with congenital Zika syndrome. Journal of Neuropathology and Experimental Neurology [Internet]. 2018;**77**(3): 193-198. DOI: 10.1093/jnen/nlx116

[61] Tan SK, Sahoo MK, Milligan SB, Taylor N, Pinsky BA. Stability of Zika virus in urine: Specimen processing considerations and implications for the detection of RNA targets in urine. Journal of Virological Methods [Internet]. 2017;**248**:66-70. DOI: 10.1016/j.jviromet.2017.04.018

[62] Johnson BW, Kosoy O, Martin DA, Noga AJ, Russell BJ, Johnson AA, et al. West Nile virus infection and serologic response among persons previously vaccinated against yellow fever and Japanese encephalitis viruses. Vectorborne and Zoonotic Diseases [Internet]. 2005;**5**(2):137-145. DOI: 10.1089/vbz.2005.5.137

[63] Mohammed H, Tomashek KM, Stramer SL, Hunsperger E. Prevalence of anti-dengue immunoglobulin G antibodies among American Red Cross blood donors in Puerto Rico, 2006. Transfusion [Internet]. 2012;**52**(8):1652-1656. DOI: 10.1111/j.1537-2995.2011.03492.x

[64] Prince HE, Matud JL. Estimation of dengue virus IgM persistence using regression analysis. Clinical and Vaccine Immunology [Internet]. 2011;**18**(12): 2183-2185. DOI: 10.1128/cvi.05425-11

[65] Busch MP, Kleinman SH, Tobler LH, Kamel HT, Norris PJ, Walsh I, et al. Virus and antibody dynamics in

acute west nile virus infection. The Journal of Infectious Diseases [Internet]. 2008;**198**(7):984-993. DOI: 10.1086/591467

[66] Theel ES, Hata DJ. Diagnostic testing for Zika virus: A postoutbreak update. Journal of Clinical Microbiology [Internet]. 2018 Apr;**56**(4). DOI: 10.1128/JCM.01972-17

[67] Oduyebo T, Igbinosa I, Petersen EE, Polen KND, Pillai SK, Ailes EC, et al. Update: Interim guidance for health care providers caring for pregnant women with possible Zika virus exposure—United States, July 2016. Morbidity and Mortality Weekly Report [Internet]. 2016;**65**(29):739-744. DOI: 10.15585/mmwr.mm6529e1

[68] Montoya JG. Laboratory diagnosis of *Toxoplasma gondii* infection and toxoplasmosis. The Journal of Infectious Diseases [Internet]. 2002;**185**(Suppl 1):S73-S82. DOI: 10.1086/338827

[69] Dale Carroll I, Toovey S, Van Gompel A. Dengue fever and pregnancy—A review and comment. Travel Medicine and Infectious Disease [Internet]. 2007;**5**(3):183-188. DOI: 10.1016/j.tmaid.2006.11.002

[70] Tunkel AR, Glaser CA, Bloch KC, Sejvar JJ, Marra CM, Roos KL, et al. The management of encephalitis: Clinical practice guidelines by the Infectious Diseases Society of America. Clinical Infectious Diseases [Internet]. 2008;**47**(3):303-327. DOI: 10.1086/589747

[71] Cordeiro MT, Pena LJ, Brito CA, Gil LH, Marques ET. Positive IgM for Zika virus in the cerebrospinal fluid of 30 neonates with microcephaly in Brazil. Lancet [Internet]. 2016;**387**(10030):1811-1812. DOI: 10.1016/s0140-6736(16)30253-7

[72] Roehrig JT, Hombach J, Barrett ADT. Guidelines for plaque-reduction neutralization testing of human antibodies to dengue viruses. Viral Immunology [Internet]. 2008;**21**(2):123-132. DOI: 10.1089/vim.2008.0007

[73] Puschnik A, Lau L, Cromwell EA, Balmaseda A, Zompi S, Harris E. Correlation between dengue-specific neutralizing antibodies and serum avidity in primary and secondary dengue virus 3 natural infections in humans. PLoS Neglected Tropical Diseases [Internet]. 2013;**7**(6):e2274. DOI: 10.1371/journal.pntd.0002274

[74] Oladapo OT, Souza JP, De Mucio B, de León RGP, Perea W, Gülmezoglu AM, et al. WHO interim guidance on pregnancy management in the context of Zika virus infection. Lancet glob. Health [Internet]. 2016;**4**(8):e510-e511. DOI: 10.1016/S2214-109X(16)30098-5

[75] Moore CA, Erin Staples J, Dobyns WB, Pessoa A, Ventura CV, da Fonseca EB, et al. Characterizing the pattern of anomalies in congenital Zika syndrome for pediatric clinicians. JAMA Pediatrics [Internet]. 2017;**171**(3):288. DOI: 10.1001/jamapediatrics.2016.3982

[76] Brasil P, Pereira JP Jr, Moreira ME, Ribeiro Nogueira RM, Damasceno L, Wakimoto M, et al. Zika virus infection in pregnant women in Rio de Janeiro. New England Journal of Medicine [Internet]. 2016;**375**(24):2321-2334. DOI: 10.1056/NEJMoa1602412

[77] Zin AA, Tsui I, Rossetto J, Vasconcelos Z, Adachi K, Valderramos S, et al. Screening criteria for ophthalmic manifestations of congenital Zika virus infection. JAMA Pediatrics [Internet]. 2017;**171**(9):847. DOI: 10.1001/jamapediatrics.2017.1474

[78] Website [Internet]. Division of vector-borne diseases. Revised diagnostic testing for Zika, chikungunya, and dengue viruses in US Public Health Laboratories. Cdc

[Internet]. 2014;**1**:1-5. Available from: http://stacks.cdc.gov/view/cdc/38149 [cited: 09 September 2018]

[79] Website [Internet]. Oladapo OT, Souza JP, De Mucio B, de León RGP, Perea W, Gülmezoglu AM, WHO Guideline Development Group. WHO interim guidance on pregnancy management in the context of Zika virus infection. Lancet Global Health [Internet]. 2016;**4**(8):e510-e511. DOI: 10.1016/S2214-109X(16)30098-5 [cited: 09 September 2018]

[80] Levine D, Jani JC, Castro-Aragon I, Cannie M. How does imaging of congenital Zika compare with imaging of other TORCH infections? Radiology [Internet]. 2017;**285**(3):744-761. DOI: 10.1148/radiol.2017171238

[81] De Santis M, De Luca C, Mappa I, Spagnuolo T, Licameli A, Straface G, et al. Syphilis infection during pregnancy: Fetal risks and clinical management. Infectious Diseases in Obstetrics and Gynecology [Internet]. 2012;**2012**:1-5. DOI: 10.1155/2012/430585

[82] Kang JH. Febrile illness with skin rashes. Infection & Chemotherapy [Internet]. 2015;**47**(3):155. DOI: 10.3947/ic.2015.47.3.155

[83] Al-Khan A, Caligiuri A, Apuzzio J. Parvovirus B-19 infection during pregnancy. Infectious Diseases in Obstetrics and Gynecology [Internet]. 2003;**11**(3):175-179. DOI: 10.1080/10647440300025518

[84] Lamont RF, Sobel JD, Carrington D, Mazaki-Tovi S, Kusanovic JP, Vaisbuch E, et al. Varicella-zoster virus (chickenpox) infection in pregnancy. BJOG: An International Journal of Obstetrics & Gynaecology [Internet]. 2011;**118**(10):1155-1162. DOI: 10.1111/j.1471-0528.2011.02983.x

[85] Picone O, Simon I, Benachi A, Brunelle F, Sonigo P. Comparison between ultrasound and magnetic resonance imaging in assessment of fetal cytomegalovirus infection. Prenatal Diagnosis [Internet]. 2008;**28**(8): 753-758. DOI: 10.1002/pd.2037

[86] Soares de Oliveira-Szejnfeld P, Levine D, Melo AS de O, Amorim MMR, Batista AGM, Chimelli L, et al. Congenital brain abnormalities and Zika virus: What the radiologist can expect to see prenatally and postnatally. Radiology [Internet]. 2016;**281**(1):203-218. DOI: 10.1148/radiol.2016161584

[87] Website [Internet]. TORCH syndrome. Lancet [Internet]. 1998;**19**(2):194-203. Available from: http://eurheartj.oxfordjournals.org/cgi/doi/10.1053/euhj.1997.0817 [cited: 23 September 2018]

[88] Nahmias AJ, Walls KW, Stewart JA, Herrmann KL, Flynt WJ. The ToRCH complex-perinatal infections associated with toxoplasma and rubella, cytomegol- and herpes simplex viruses. Pediatric Research [Internet]. 1971;**5**(8):405-406. DOI: 10.1203/00006450-197108000-00144

[89] Hampton MM. Congenital toxoplasmosis: A review. Neonatal Network [Internet]. 2015;**34**(5):274-278. DOI: 10.1891/0730-0832.34.5.274

Chapter 7

Semi-Solid Phase Assay for the Alternative Complement Pathway Activity Assessment (AP_{100})

Kheir Eddine Kerboua and Kamal Djenouhat

Abstract

Since the introduction of the most expensive drug in the world (Eculizumab) in the therapeutic arsenal of many diseases involving the alternative complement pathway (ACP) in their pathophysiology, the unmet need to perform simple ACP assays affordable for all countries has become one of the major challenges of the contemporary medicine. The assay currently used is AH_{50}, despite it still challenging for several laboratories. This educational chapter consists on a detail protocol of standardized hemolytic assay AP_{100} and aims to help clinical laboratories over the world and especially those of the developing and low incomes countries to perform it. The procedure is essentially the same as for the timed lysis assay and dilution methods (AP_{50}) except the concentration of ACP buffer and the chicken erythrocyte density used to make the gels. In clinical field, AP_{100} has at least nine applications in disease diagnosis and follow-up. AP_{100} has many advantages over the AH_{50} as it is more reliable for the Eculizumab monitoring and more practical with a purpose to be stored and transported for several weeks. AP_{100} is a portable and easy to use device both at the bedside and in the companion medical care.

Keywords: alternative complement pathway, AP_{100}, medicine, simple assay, low income countries

1. Introduction

Complement system is the pillar of the immune system by its dual role in homeostasis and disease. It is the first line of the innate immunity and augments adaptive immunity. Indeed, complement acts as a rapid and efficient immune surveillance system that has distinct effects on healthy and altered host cells and foreign intruders through a complex cascade of proteases [1]. Activation of the pathway occurs through three primary pathways: classical, lectin, and alternative pathways. Instead to the other pathways and in addition to properdin as the initiating molecule, the alternative complement pathway (ACP) is activated via a low level of constitutive spontaneous hydrolysis of C3 in a process known as tick-over. Importantly, thanks to its amplification loop, ACP plays a major role for the final effect of initial specific activation of the classical and lectin complement pathways and contributed to 80–90% of any C5 activation regardless the initiating pathway [2]. Interestingly, ACP has been shown to play a particularly important role in preclinical disease models [3].

Disease presentation	ACP proteins involved
Immune complex (IC) diseases, SLE with and without glomerulonephritis (GN).	Most of ACP proteins trigger tissue injury because ACP constitutes the amplification loop of the classical pathway
Severe and often repeated pyogenic infections.	Deficiency of C3 either primary, or secondary to fH or fI deficiency, or the presence of C3 nephritic factors. Can also occur with classical component deficiency.
Mesangiocapillary glomerulonephritis (MCGN) or membranoproliferative glomerulonephritis (MPGN)	C3 deficiency classically caused by C3 nephritic factors, also fH deficiency
Neisserial infections, usually recurrent meningococcal infections. Occasionally disseminated *N. gonorrhea* infections.	Terminal component (C5–8) deficiencies; the association is less strong in patients with C9 deficiency. Patients are usually otherwise healthy.
Severe, occasionally repeated, *N. meningitidis* infections in males	Properdin deficiency or Factor D deficiency (very rare, not sex linked)
Paroxysmal nocturnal hemoglobinuria	Rare acquired clonal disorder of hemopoietic cells which lack glycosylphosphatidylinositol (GPI) membrane anchored Complement control proteins [(DAF) And CD59].

Table 1.
Clinical presentations of ACP proteins acquired or genetic abnormalities.

The ACP functional assessment constitutes an unmet need in medicine and applied research fields as the health valorization of bio-molecules extracted from nature. In clinical field it has at least nine applications in disease diagnosis and follow-up. For instance in therapy monitoring, it allows to screen patients responders to the complement blockers like Eculizumab, a patient with abolished activity means that he has no C5 mutation and is considered as eligible to this therapy. Furthermore, ACP activity makes possible to assess drug effectiveness at the plasmatic level [4]. Moreover, by evaluating ACP function we can predict and avoid immune-complex diseases flares and end organs damages as in systemic lupus erythematosus (SLE).

For the disease diagnosis, several international consensuses include functional hypocomplementemia and ACP abnormalities as a diagnostic criterion:

1. Kidney diseases resulting from abnormal control of ACP especially atypical hemolytic uremic syndrome (aHUS), C3 glomerulonephritis (C3GN), and dense-deposit disease (DDD), as well as atypical postinfectious glomerulonephritis [5].

2. Hypocomplementemia by a hemolytic assay constitutes one point in the diagnosis score of EULAR/ACR Lupus Classification Criteria 2017 and useful marker for evaluating SLE renal disease activity and outcomes [6].

3. The clinical hallmark of paroxysmal nocturnal hemoglobinuria (PNH) is the chronic intravascular hemolysis that is a consequence of unregulated activation of ACP [7].

4. Individuals deficient in components of the alternative and terminal complement pathways are highly predisposed to invasive, often recurrent meningococcal infections [8]. The most frequent bacterial meningitides related to complement proteins deficiencies are dues to factor B, factor D and membrane complex attack proteins deficiency.

5. Hypocomplementemic hypersensitivity reactions to synthetic hemodialysis membrane at the origin of cardiovascular complication, the most frequent and life-threatening complication in hemodialysis [9].

6. Several clinical presentations linked to ACP abnormalities as resumed in **Table 1** [2].

7. ACP is particularly considered in sepsis, due to its uncontrolled amplification in sepsis conditions [10].

2. Protocol

The procedure is essentially the same as for the ACP by kinetic fluid phase assay except the concentration of AP buffer and the chicken erythrocyte density used to make the gels [11]. A value of 100% of plasma ACP function should be defined by the pooled normal human plasmas (NHP standard), prepared from a total of 100 healthy individuals separate for each sex.

2.1 Samples

Patient serum samples for the functional hemolytic assays need to be fresh, that is serum should be separated on the day of venepuncture and used the same day, or stored at −80°C. This is probably the single most difficult, yet important, step because if the cold chain is broken, the results become impossible to interpret correctly [12]. Whole blood, with ethylenediaminetetraacetic acid (EDTA) as the anticoagulant can also be used.

Chicken red blood cells (CRBC) was collected in tubes containing Alsever or 20% (v/v) acid citrate dextrose (ACD) and stored at 4°C.

2.2 Buffers and other reagents

1. Phosphate-buffered saline (PBS) contains: 137 mM NaCl, 2.7 mM KCl, 1.5 mM KH_2PO_4, 8.1 mM Na_2HPO. Sodium azide can be added if required.

2. Alternative complement pathway buffer (ACP buffer) is PBS containing 100 mM ethylene glycol-bis(β-aminoethyl ether)N,N′,N′-tetraacetic acid (EGTA) and 7 mM $MgCl_2$; resulting in chelation of Ca^{2+}, but not Mg^{2+}, and providing additional Mg^{2+} [13]. This prevents complement activation via the classical pathway and facilitates complement activation via the ACP in agarose gels. ACP buffer/gelatin contains 1 g/L gelatin and is used when protein concentrations are low.

3. Agarose: use an agarose that has a low melting point as plates are easier to pour.

2.3 Procedure

1. Before starting, in a sterile tube put 9.8 mL of 1×ACP buffer at 56°C. Also let warming some Petri dish (or rectangular plate) in the 56°C incubator.

2. Preparation of chicken erythrocytes (CE) for ACP assay

a. Under aseptic conditions, remove 100 μl packed CE from stock chicken blood stored in Alsever's solution.

b. Wash twice in ACP buffer.

c. Resuspend in the same buffer to the required concentration for assay by 2.1 mL of ACP buffer and put the cell suspension in a water bath at a temperature of 46°C. The concentration of CE can be calculated by lysing 0.1 mL of the stock CE in 2.9 mL H_2O and measuring the absorbance of the supernatant at 412 nm.

3. Melt 2% agarose stock (most conveniently in a microwave oven although immersion in boiling water will suffice) and, using a warm pipet, pipet 12.25 mL aliquots into universal containers (one for each gel), keep at 56°C.

4. Add the 9.8 mL warmed 1×ACP (56°C) to each bottle of melted agarose. Mix well.

   Transfer one bottle with diluted agarose to 45°C and allow to cool to this tempe rature.

5. To ease pouring the gel, place the warmed Petri dish (or rectangular plate) on level tray. Mix carefully and quickly.

6. Pour the mixture evenly onto the level plate. The mixture should go to the mid edges of the Petri dish/rectangular plate. Remove bubbles by touching them with a pipette tip or gloved finger.

7. Cool plate to 4°C, punch holes using a Pasteur pipette upside down at least 1 cm apart.

8. Fill wells with a measured serum volumes (30 μL). Include a normal human plasma (NHP) standard, and NHP diluted 1/2 and 1/4 on each individual dish/plate. The size of rings depends on factors such as gel thickness, and NHP standards are needed on each dish.

9. Incubate overnight at 4°C, examine Petri dish before transferring to 37°C. Incubate at 37°C for 1–2 h. Incubate either overnight at 22°C and 2 h at 37°C.

10. Measure the diameters of the rings of lysis and calculate the areas. Areas of lysis can be read after photography, or after making direct photographic prints, but this is optional.

11. Standard curve: A crude standard curve is drawn by plotting % concentration NHP *vs.* area of lysis (diameter squared can also be used). This allows calculation of % normal activity in test samples.

3. Representative results

After incubation, the ACP activity is calculated after measuring the diameter using the vernier caliper or taking the ACP dish in photo and measure each surface well by *Image J*® (**Figure 1**). Area surfaces are calculated by the formula $S = \pi^* r^2$,

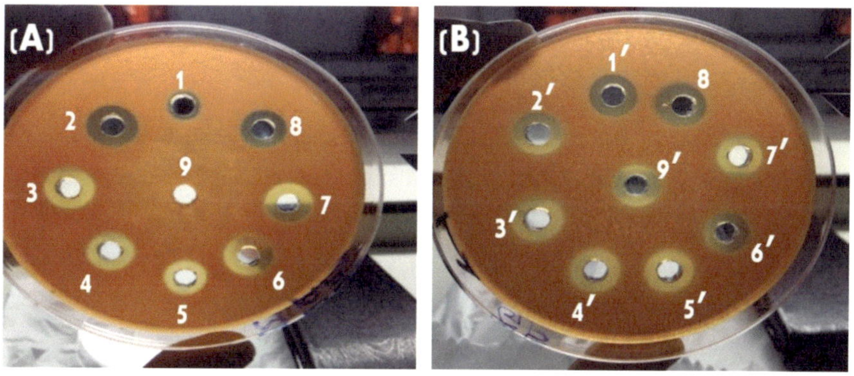

Figure 1.
ACP dishes for different sera from patients and pooled normal plasmas NHP. (A) Rings size of well 1 and 9 are found in a complete ACP blockade by Eculizumab or total deficiency of one of the ACP proteins. The remaining wells show different plasmatic levels of ACP. (B) Sera tested in this ACP dish are within normal range but vary according to the clinical presentation.

with r = d/2. (r: radius, d: diameter, s: surface). The same procedure is applied to the different NHP dilutions included to establish the calibration curve. This is because diluting NHP means that all ACP components are diluted equally. As example of interpretation, the more the ring is small the more the effectiveness of Eculizumab is good. The absence of reaction in a well signifies a complete ACP blockage, either by a drug or by a pathological process.

3.1 Materials

1. Chicken erythrocytes (CE) conserved in Alsever's solution.

2. Phosphate-buffered saline (PBS).

3. Veronal buffered saline (VBS).

4. Gelatin veronal buffer (GVB).

5. AP buffer is GVB containing 5 mM Mg and 5 mM ethylene glycol bis[β-aminoethylether]N,N′-tetraacetic acid (EGTA, E4378 Sigma).

6. N-saline is 9 g NaCl dissolved in 1 L H_2O.

7. Barbitone buffer is made by mixing 0.1 M solutions of sodium barbitone and barbituric acid to obtain the target pH and adjusting volume to obtain required final molarity.

3.2 Equipment

1. Suitable Petri dishes or glass plates. Size depends on the number of samples. Volumes can be adjusted so that the final depth of the gel is ~1–1.5 mm.

2. 56°C incubator and dishes/plate warmer (water bath can also be used).

3. 46°C water bath.

4. Level table.

5. 37°C incubator.

6. Pasteur pipette as well cutters to produce holes approximately 5 mm diameter. If single holes are to be punched, a grid should be placed under the gel so that the holes can be distributed evenly.

7. Refrigerated centrifuge.

8. Electronic balance.

9. pH meter/temperature probe.

4. Discussion

Since Thomas A. E. Platts-Mills and Kimishige Ishizaka have discovered that fresh normal human serum in EGTA buffer was found to cause >90% hemolysis of unsensitized rabbit red blood cells (RaRBC) [11], simple timed lysis assay and dilution methods called AP_{50} (Alternative Pathway 50) was performed to quantify hemolytic complement activity in human serum [13–15]. This reaction requires C3, factors B and D, and Mg^{++} ions to form the C3 convertase (C3bBb) [14, 15]. In AP_{50}, ACP is activated and measured by virtue of RaRBC decreased sialic acid content in addition to the blockade of the classical pathway activation by chelation of calcium by EGTA [16]. Nonetheless, AH_{50} requires a lot of material like spectrophotometer and consumable test tubes and highly skillful personal. To adapt this assay to simple labs, a semi-solid phase assay was proposed to measure ACP activity by using chicken erythrocytes incorporated in agarose gel called AP_{100}. Cell membranes of chicken erythrocytes have the same properties as those of the

	Fluid phase assay (AH_{50})	**Semi-solid phase assay (AP_{100})**
Number of samples	<5 per day	Several samples series per day
ACP Activators	Red cells	Red cells + agarose particles
End point assessment	Optical density by Spectrophotometer	Diameter measurement directly
Laboratory work flow steps	+++	Less
Time consuming	+++	Less
Storage	No	15–21 days
Transport	No	Yes (respecting the cold chain)
Global cost	+++	Less
Assay related error	+++	Less
Manipulator linked error	+++	Less
Others		• More suitable complement related diseases screening • Do not need a deep immunotechnology handling • Storage of utilized plates for *intra* and *inter*-laboratory comparisons

Table 2.
Hemolytic agarose assay features in comparison to tube hemolytic assay.

rabbit in terms of cell surface charge density of sialic acid whereas they are more robust than rabbit because they have nuclei. AP_{100} assay has many advantages over the AH_{50} one (**Table 2**). For example, AP_{100} is more reliable for the eculizumab monitoring as shown in the article we have recently reviewed [17]. Moreover, AH_{50} still challenging in low incomes country laboratories that cannot equip their hospital and research laboratories to perform this assay. To overcome this roadblock, it was proceeded to render this method more practical to each laboratory with a purpose to be stored and transported for several weeks. This includes a hemolytic agarose dishes/plates; a special adaptation to prevent complement activation via the classical pathway and facilitate complement activation via the alternative one in agarose gels.

To the best of our knowledge there were no laboratory performing that assays in developing countries especially in Africa and no educational chapter is available to explain it. Once its optimization in each of these countries laboratories was performed, thanks to ACP dishes/plates, we are expecting to empower doctors' decision making process and improve quality of patients' management and therapy follow-up. Therefore this portable and easy to use device even at the bedside of the patient and in the companion do not necessitate any equipped laboratory and may facilitate prospective analysis and disease screening in large populations. With enlargement of the ACP disease spectrum necessitated complement blockade, AP_{100} should be considered by clinical laboratory scientist especially to analyze sets of samples at once.

Acknowledgements

I am very thankful to my all students; Ahlem Lamnai, Souad M'hamedi, Saadia Benmadi, Wasila Aeid.

Disclosures

The author has nothing to disclose.

A.Appendix

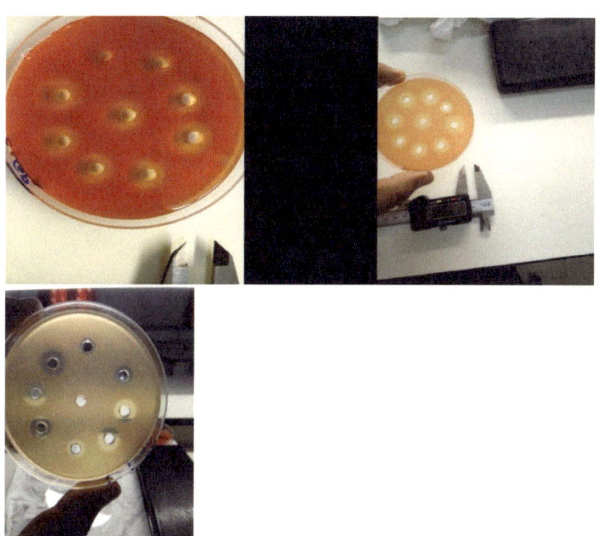

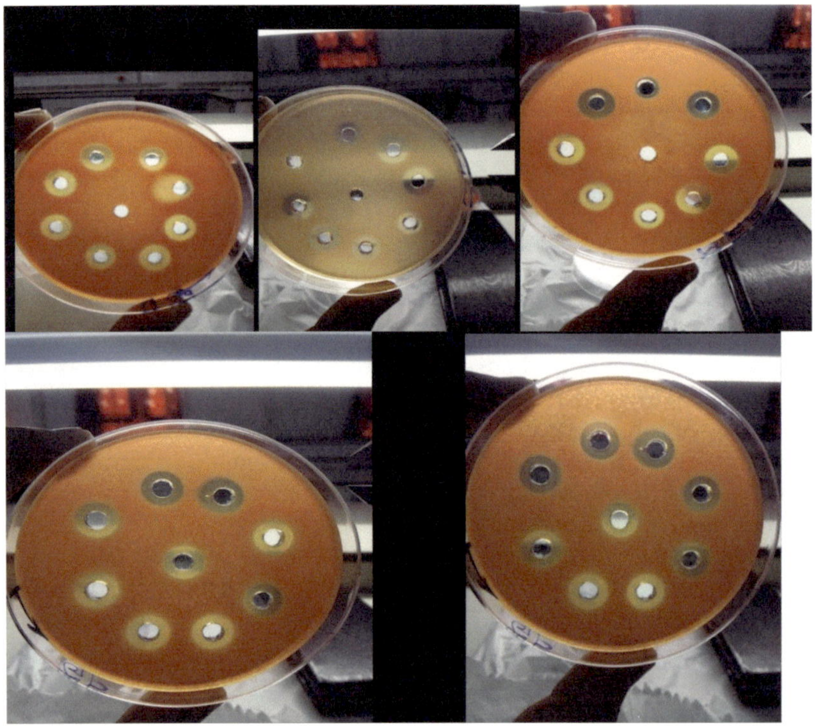

Author details

Kheir Eddine Kerboua[1*] and Kamal Djenouhat[2]

1 Laboratory of Immunology, Faculty of Medicine, University of Kasdi Merbah Ouargla, Algeria

2 Laboratory of Immunology, Faculty of Medicine, University of Algiers I, Algeria

*Address all correspondence to: k.k.eddine@gmail.com

References

[1] Ricklin D, Hajishengallis G, Yang K, Lambris JD. Complement: A key system for immune surveillance and homeostasis. Nature Immunology. 2010;**11**(9):785-797

[2] Shih AR, Murali MR. Laboratory tests for disorders of complement and complement regulatory proteins. American Journal of Hematology. 2015;**90**(12):1180-1186

[3] Holers VM. The spectrum of complement alternative pathway-mediated diseases. Immunological Reviews. 2008;**223**:300-316

[4] Kerboua K et al. C3:CH50 ratio as a proposed composite marker for eculizumab monitoring in atypical hemolytic uremic syndrome: Preliminary results. Journal of Immunoassay and Immunochemistry. 2017;**38**(2):178-189

[5] Angioi A, Fervenza FC, Sethi S, Zhang Y, Smith RJ, Murray D, et al. Diagnosis of complement alternative pathway disorders. Kidney International. 2016;**89**(2):278-288

[6] Song D, Guo WY, Wang FM, Li YZ, Song Y, Yu F, et al. Complement alternative pathway′s activation in patients with lupus nephritis. The American Journal of the Medical Sciences. 2017;**353**(3):247-257

[7] Lindorfer MA, Pawluczkowycz AW, Peek EM, Hickman K, Taylor RP, Parker CJ. A novel approach to preventing the hemolysis of paroxysmal nocturnal hemoglobinuria: Both complement-mediated cytolysis and C3 deposition are blocked by a monoclonal antibody specific for the alternative pathway of complement. Blood. 2010;**115**(11):2283-2289

[8] Lewis LA, Ram S. Meningococcal disease and the complement system. Virulence. 2014;**5**(1):98-126

[9] Rodríguez-Sanz A, Sánchez-Villanueva R, Domínguez-Ortega J, Fiandor A-M, Ruiz M-P, Trocoli F, et al. Mechanisms involved in hypersensitivity reactions to polysulfone hemodialysis membranes. Artificial Organs. 2017;**41**(11):285-295. DOI: 10.1111/aor.12954

[10] Markiewski MM, DeAngelis RA, Lambris JD. Complexity of complement activation in sepsis. Journal of Cellular and Molecular Medicine. 2008 Dec;**12**(6a):2245-2254

[11] Pangburn MK. The alternative pathway. In: Ross G, editor. Immunobiology of the Complement System. San Diego: Academic Press; 1981. pp. 45-62

[12] Pangburn MK. Alternative pathway of complement. Methods in Enzymology. 1988;**162**:648-649

[13] Platts-Mills TAE, Ishizaka K. Activation of the alternate pathway of human complement by rabbit cells. Journal of Immunology. 1974;**113**(1):348-358

[14] Servais G, Walmagh J, Duchateau J. Simple quantitative haemolytic microassay for determination of complement alternative pathway activation (AP50). Journal of Immunological Methods. 1991;**140**(1):93-100

[15] Kazatchkine M, Hauptmann G, Nydegger U. Dosage de l'activité fonctionnelle de la voie alterne VAH50. In: Inserm, editor. Techniques en Immunologic. Techniques du Complement. Paris: Inserm; 1985. p. 34

[16] Kazatchkine MD, Nydegger UE. The human alternative complement pathway: Biology and immunopathology of activation and regulation. Progress in Allergy. 1982;**130**:193

[17] Puissant-Lubrano B, Puissochet S, Congy-Jolivet N, Chauveau D, Decramer S, Garnier A, et al. Alternative complement pathway hemolytic assays reveal incomplete complement blockade in patients treated with eculizumab. Clinical Immunology. 2017;**183**:1-7

Chapter 8

Amino Acids Profiling for the Diagnosis of Metabolic Disorders

Yana Sandlers

Abstract

Inborn errors of metabolism (IEM) represent a group of inherited diseases in which genetic defect leads to the block on a metabolic pathway, resulting in a single enzyme dysfunction. As a downstream consequence of the residual or full loss of the enzymatic activity, there is an accumulation of toxic metabolites in the proximity of the metabolic block and/or a deficiency of an essential metabolic product which leads to the clinical presentation of the disease. While individually IEMs are rare, a collectively estimated incidence of metabolic inherited disorders is 1:800. The genetic basis of IEMs can involve abnormalities such as point mutations, deletions or insertions, or more complex genomic rearrangements. Categorization of IEM can be simply made on the basis of the affected metabolic network: fatty acids oxidation disorders, protein/amino acids metabolism disorders, disorders of carbohydrate metabolism, lysosomal storage diseases, peroxisomal disorders, and mitochondrial diseases. This chapter will overview amino acid metabolism-related inherited disorders and amino acid analysis for the diagnosis and routine monitoring of this category of IEMs.

Keywords: inborn error of metabolism, amino acids disorders, quantitative amino acids analysis, ion exchange chromatography, mass spectrometry

1. Introduction

Amino acids (**Figure 1**) play multiple important roles in our body: they are basic structural protein units and precursors of neurotransmitters, porphyrins, and nitric oxide. Furthermore, amino acids derived from the dietary proteins serve as energy source since while catabolized in our body, amino acids form organic acids that can replenish Krebs cycle and ammonia that eliminates through urea cycle [1].

Amino acids disorders (also called aminoacidopathies) are a group of inborn errors of metabolism diseases, caused by the inherited defects in pathways involved in amino acids metabolism. All primary amino acids disorders (**Table 1**) follow an autosomal recessive mode of inheritance which means that the mutation caused a metabolic block is present in the genetic material of both parents. As a result of mutation, the inherited defect is reflected downstream as a lack or a partial biological activity of enzymes involved in amino acids metabolism. Consequently, some substrates in these pathways accumulate or are diverted into alternative pathways. Therefore, amino acids disorders are biochemically characterized by abnormal levels of single or several amino acids and their downstream plasma and/or urine metabolites (**Tables 2–6**). Amino acid disorders are presented with variable and often nonspecific clinical symptoms. In conjunction with medical support, these disorders are managed by nutritional restrictions, supplements and medical

IntechOpen

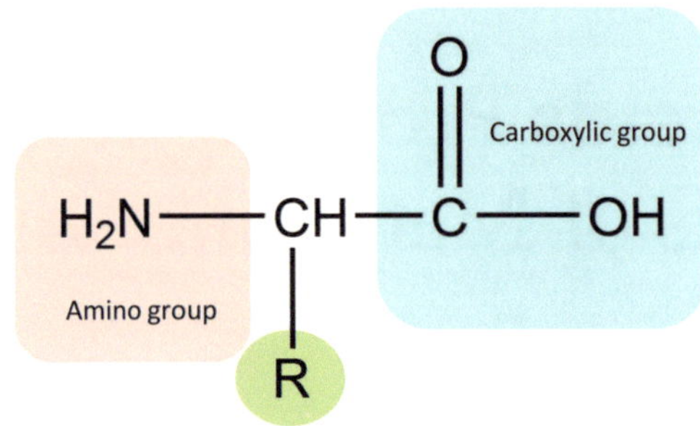

Figure 1.
The general structure of amino acids consist of an amino group, a carboxylic group and a variable R side chain that has a major effect on solubility and polarity.

foods that limit consumption of an offending amino acid or in some cases protein consumption. It is important therefore routinely perform amino acids' analysis to monitor dietary treatment outcomes in already diagnosed individuals. In the next chapters, primary amino acids disorders are reviewed and quantitative amino acid analysis in clinical settings is discussed.

1.1 Phenylketonuria

Phenylketonuria (commonly known as PKU, incidence 1 in 13,500–19,000 births in the United States [2]) is an inherited disorder of phenylalanine metabolism characterized by phenylalanine hydroxylase deficiency (**Figure 1**). The enzyme catalyzes the conversion of phenylalanine to tyrosine in the presence of tetrahydrobiopterin (BH4) as a cofactor. Based on plasma phenylalanine level, PKU is categorized by severe (Phe > 1200 μmol/L), mild (Phe = 600–1200 μmol/L) and hyperphenylalaninemia (above the normal cut off but below 600 μmol/L) phenotypes. Clinically PKU can be presented with growth failure, global developmental delay, severe intellectual disabilities and other severe symptoms. During pregnancy, elevated levels of phenylalanine have teratogenic effects on the developing fetus [3] and the condition is recognized as maternal PKU. Phenylalanine accumulation is also seen in defects of biopterin cofactor biosynthesis and regeneration [4] (**Table 1**). Nutritional management of PKU targets excessive accumulation of phenylalanine by restriction of natural protein intake in combination with the use of phenylalanine-free protein substitutes.

1.2 Disorders of tyrosine metabolism

Tyrosine metabolic pathway consists of five enzymatic reactions taking place mainly in hepatocytes and renal proximal tubules. Tyrosinemia I is the most severe inherited disorder of tyrosine metabolism caused by a deficiency of fumarylacetoacetate hydrolase, the last enzyme in the tyrosine catabolic pathway. The disorder has a high incidence in French Canadian ethnicity [5] and involves hepatorenal dysfunction. Tyrosinemia II is caused by a deficiency of the hepatic tyrosine aminotransferase and manifested by mental retardation and other severe symptoms [6]. A deficiency in the activity of 4-hydroxyphenylpyruvate dioxygenase leads to tyrosinemia III, a rare disorder characterized by mild mental retardation and/or convulsions [7, 8]. All three disorders biochemically characterized by high levels of plasma tyrosine (hypertyrosinemia) and urine excretion of downstream tyrosine metabolites (**Table 1**). Elevated plasma tyrosine can also be seen due to vitamin-C responsive transient tyrosinemia during the neonatal period (**Figure 2**).

Aromatic amino acids disorders			
Disorder name	**Amino acid involved**	**Enzyme or transport defect**	**Additional biomarkers**
PKU classical	Phe (B) high	PAH	Phe: Tyr ratio (B), phenylpyruvic, phenyllactic and 2-hydroxyphenylacetic acids (U)
Defect of biopterin cofactor biosynthesis	Phe (B) high	GTPCH	Low biopterin, neopterin (U)
Defect of biopterin cofactor biosynthesis	Phe (B) high	PTPS	Low biopterin, high neopterin (U)
Defect of biopterin cofactor regeneration	Phe (B) high	PCBD1	High neopterin and primapterin (U)
Defect of biopterin cofactor regeneration	Phe (B) high	DHPR	High biopterin (U) and low DHPR activity in dried blood spots
Tyrosinemia I	Tyr (B) high	FAH	Succinylacetone (DBS, U), 4-hydroxy-phenylpyruvic, 4-hydroxy-phenyllactic acids (U)
Tyrosinemia II	Tyr (B) high	TAT	4-hydroxyphenylpyruvic, 4-hydroxy-phenyllactic acids (U)
Tyrosinemia III	Tyr (B) high	HPPD	4-hydroxyphenylpyruvic, 4-hydroxyphenyllactic acids (U)

PAH, phenylalanine hydroxylase; GTPCH, GTP cyclohydrolase; PTPS, 6-pyrovoyltetrahydropterin synthase; PCBD1, pterin-4a-carbinolamine dehydratase; DHPR, dihydropterin reductase; FAH, fumarylacetoacetate hydrolase; TAT, tyrosine aminotransferase; HPPD, 4-hydroxyphenylpyruvate dioxygenase.

Table 1.
Laboratory findings in aromatic amino acids disorders.

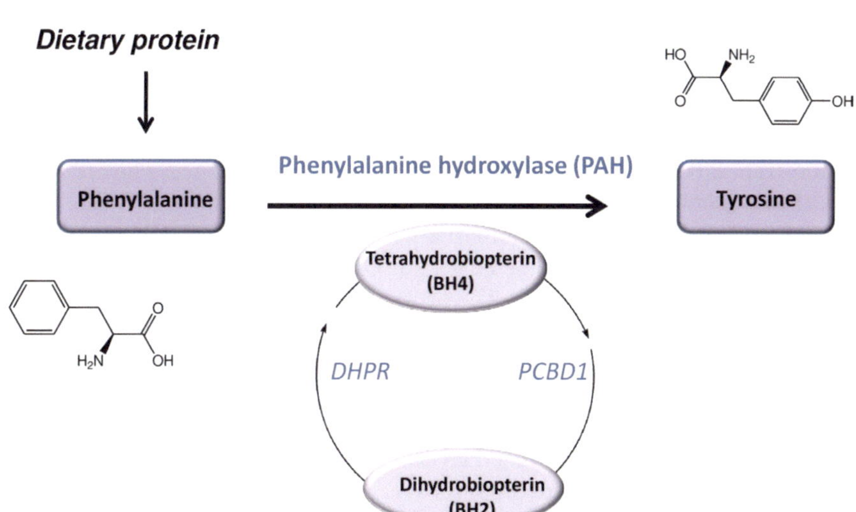

Figure 2.
Reaction catalyzed by phenylalanine hydroxylase. Tetrahydrobiopterin (BH_4) is a co-factor of PAH. DHPR, dihydropteridine reductase; PCBD1, pterin-4-α-carbinolamine dehydratase.

1.3 Maple syrup urine disease

Maple syrup urine disease is a disorder of branch chain amino acids metabolism caused by a deficiency of branched-chain α-keto acid dehydrogenase complex. MSUD is presented with five clinical phenotypes on the basis of the age at onset, the severity of symptoms and response to thiamine supplementation [9]. MSUD characterized biochemically by elevated plasma branched-chain amino acids (leucine, isoleucine, valine, allo-isoleucine) and their abnormal ratio (normal ratio is valine:is oleucine:leucine/3.5:1:2). The disease is managed by dietary leucine restriction, thus all branch chain amino acids and allo-isoleucine are routinely monitored. The classic MSUD is the most severe form of the disease characterized by no or very low residual enzyme activity and clinically manifested by developmental and neurological delays, encephalopathy, feeding problems, and a characteristic maple syrup odor in urine.

Disorders branched-chain amino acids

Disorder name	Amino acid involved	Enzyme or transport defect	Additional biomarkers
MSUD	BCAA (B) high, allo-Ile (B) high	BCKDC	Plasma ratio of Val:Ile:Leu (3.5:1:2), branch chain 2-ketoacids and 2-hydroxyacids (U)

BCKDC, branched-chain ketoacid dehydrogenase complex.

Table 2.
Laboratory findings in MSUD.

1.4 Urea cycle disorders

During protein catabolism, amino acids' carbon skeleton is metabolized to gluconeogenic and/or ketogenic precursors whereas nitrogen group is converted to ammonia through the deamination process. Toxic ammonia derived from amino acids and other metabolic sources is entering the urea cycle and further is converted to the readily excreted and nontoxic urea. The cycle takes place in the liver and a deficiency of any enzymes or transporters involved in the urea cycle can cause ammonia accumulation (hyperammonemia) which has a highly toxic effect on the central nervous system. The overall estimated incidence of urea cycle disorders is 1:8000. All urea cycle disorders have an autosomal recessive inheritance, with the exception of ornithine-transcarbamylase deficiency (OTCD), which is X-linked. Plasma citrulline is a key amino acid in the biochemical diagnosis of urea cycle defects (**Table 3**).

Hyperornithinemia-hyperammonemia-homocitrullinuria (HHH) syndrome is caused by the mutations in the SLC25A15 or ORNT1 gene which result in the deficiency of ornithine translocase. The protein transports ornithine, lysine, and arginine across the inner mitochondrial membrane in peripheral tissues and pericentral hepatocytes. ORC1 deficiency reduces the availability of mitochondrial ornithine, which leads to the ornithine increase in the cytosol (hyperornithinemia). In the liver, since the mitochondrial ornithine is a required substrate for ornithine transcarbamylase (OTC), the reduced level of mitochondrial ornithine slows down flux through the urea cycle (**Figure 3**). As a result of the reduced capacity of the urea cycle, ammonia and carbamoyl-phosphate levels increase (hyperammonemia). At the same time, an excess of carbamoyl-phosphate is diverted to react with lysine to form homocitrulline (homocitrullinuria) or enters in the pyrimidine pathway, to form orotic acid which is later excreted in urine. Similarly, as for other urea cycle disorders, early diagnosis in infancy may improve the clinical outcome of HHH.

Urea cycle disorders			
Disorder name	**Amino acid involved**	**Enzyme or transport defect**	**Additional biomarkers**
Citrullinimia I	Cit (B) markedly elevated	ASS	Hyperammonemia, orotic acid (U), can be accompanied by high glutamine and alanine (B)
Citrullinimia II/citrin deficiency	Cit (B) moderate high, Met, Lys (B)	Asp/Glu mitochondrial exchanger	Hyperammonemia, orotic acid (U). Citrulline is moderately elevated
CPS-I deficiency	Cit (B) low	CPS1	Can be accompanied by high glutamine and alanine (B)
Ornithine transcarbamylase deficiency	Cit (B) low	OTC	Nonspecific amino acid profile: increased glutamine, alanine and decreased ornithine, arginine (B). Orotic acid (U) markedly increased
Arginino-succinic acidemia	ASA (B), (U) elevated	ASL	Low citrulline, low arginine (B)
Argininemia	Arg (B) high	Arginase	Orotic acid (U), Normal or reduced citrulline (B)
HHH	Orn (B, U), high	ORC1	Homocitrulline (U) high
Co-factor producing *N*-acetyl glutamate synthetase deficiency	Gln (B) high	NAGS	Cit (B) low, alanine high (B)

B, blood; U, urine.

Table 3.
Laboratory findings in urea cycle disorders.

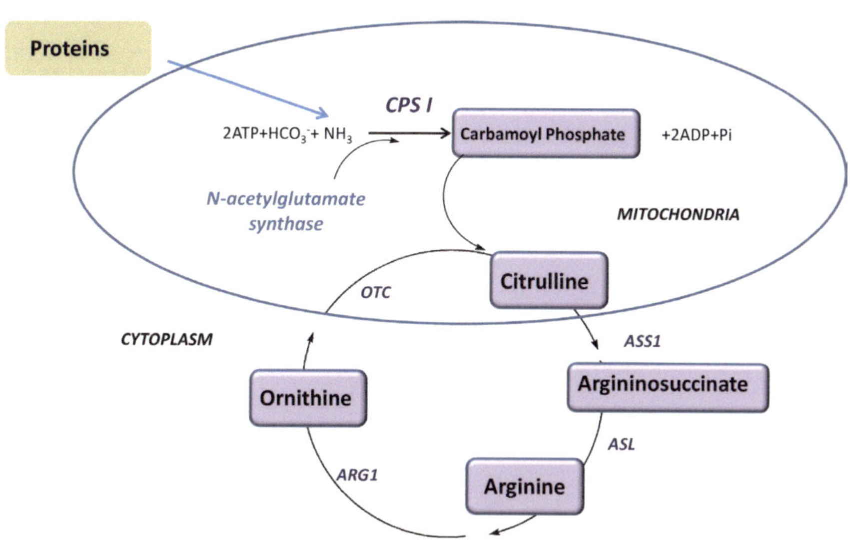

Figure 3.
Urea cycle. ARG, arginase; ASS, argininosuccinate synthase; ASL, argininosuccinate lyase; CPS1, carbamoylphosphate I synthase; ORNT 1, mitochondrial ornithine transporter; OTC, ornithine transcarbamylase.

1.5 Disorders of sulfur amino acids metabolism

Homocystinuria is a disorder of methionine metabolism (**Figure 4**). The main biochemical finding in homocystinuria is accumulation of a sulfur-containing amino acid homocysteine and its metabolites in the blood and urine. Homocysteine is formed from methionine via transmethylation. Once generated homocysteine can be irreversibly degraded via transsulfuration pathway to cysteine or remethylated back to methionine by methionine synthase. Remethylation involves a transfer of methyl group from 5-methyltetrahydrofolate to homocysteine via cobalamin (Cbl)-dependent methionine synthase (MT) and links folate cycle and homocysteine pathway. Homocysteine can also be remethylated through an additional pathway which involves liver and kidney betaine-homocysteine methyltransferase. Defects in any of these steps can result in homocystinuria. The classic homocystinuria is caused by cystathionine β-synthase (CBS) deficiency [10], a key enzyme in the trans-sulfuration pathway that converts homocysteine into cystathionine. A block at cystathionine β-synthase limits transsulfuration to the cysteine and results in both increased homocysteine and methionine, the latter caused by enhanced remethylation. The remethylation homocystinuria disorders include methylene-tetrahydrofolate reductase deficiency (MTHFR) and defects of cobalamin (Cbl) metabolism [11]. It has to be noted that methionine and not homocysteine is analyzed through the newborn screening, thus, MTHFR disorder and the cobalamin defects may not be detected because methionine level in these disorders can be normal. To increase the detection rate in cobalamin related disorders and MTHFR, some studies report a benefit of adding total homocysteine analysis to the diagnostic workflow [12]. Total homocysteine is defined as the sum of all homocysteine species in plasma/serum, including free and protein-bound forms. The measurement of total homocysteine requires an immediate separation and freezing of the collected plasma.

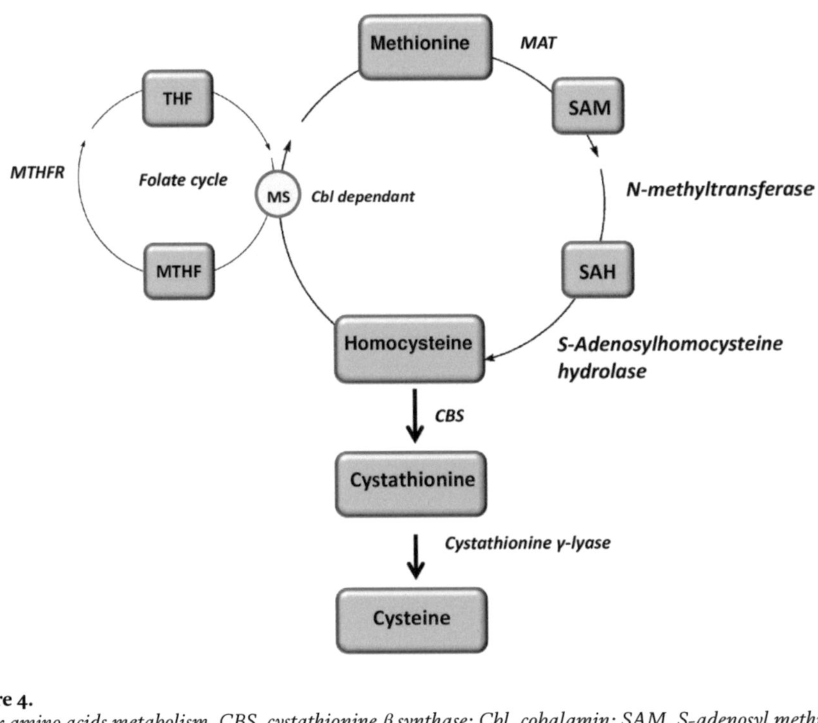

Figure 4.
Sulfur amino acids metabolism. CBS, cystathionine β synthase; Cbl, cobalamin; SAM, S-adenosyl methionine; SAH, S-adenosylhomocysteine; MAT, methionine adenosyltransferase; MS, methionine synthase; MTHFR, methylene tetrahydrofolate reductase; THF, tetrahydrofolate.

Disorders of sulfur amino acids			
Disorder name	**Amino acid involved**	**Enzyme or transport defect**	**Additional biomarkers**
Homocystenuria	Free homocystine (B, U) high	CBS	Total Hcy and methionine (B, U)
S-Adenosyl-homocysteine-hydroxylase deficiency	Met (B) high	AdoHcyas	Mildly elevated total plasma Hcy, S-adenosylhomocysteine, (B), S-adenosylmethionine (B)
Sulfate oxydase deficiency	SSC* (B, U) high	SUOX	Taurine, low cystine (B, U)
Molybden cofactor deficiency	SSC* (B, U) high	SUOX, XDH, AO	Taurine, low cystine (B, U), elevated hypoxanthine and xanthine (U), low uric acid (B)
Glycine-*N*-methyltransferase deficiency	Met (B) high	GNMT	S-adenosylmethionine (B)
Hyper-methioninemia	Met (B) high	MAT	

CBS, cystathionine beta synthase; AdoHcyas, S-adenosylhomocysteine hydroxylase; SUOX, sulfate oxidase; XDH, xanthine dehydrogenase; AO, aldehyde oxidase; GNMT, glycine-N-methyltransferase.
S-sulphocysteine may not be detectable in plasma using routine methods in sulfite oxidase and molybdenum co-factor deficiencies.

Table 4.
Laboratory findings in disorders of sulfur amino acids.

1.6 Nonketotic hyperglycemia

Nonketotic hyperglycemia (NKH) is a severe disorder of glycine metabolism. Glycine is catabolized through the four-peptide cleavage complex. P-protein, a pyridoxal phosphate-containing protein, T-protein, a protein required for the tetrahydrofolate-dependent reaction, H-protein, a protein that carries the amino-methyl intermediate and then hydrogen through the prosthetic lipoyl moiety, and L-protein, a lipoamide dehydrogenase. The disorder is so severe, that most of the affected individuals die within few months of life or survive with significant intellectual disabilities. Main laboratory findings in NKH is plasma and CSF elevated glycine.

Disorders of amino acids transport			
Disorder name	**Amino acid involved**	**Enzyme/transport defect**	**Additional biomarkers**
Cystinuria	Cystine (U) elevated	Cystine and dibasic amino acids in GI tract and renal tubule	Lysine, ornithine, arginine increase (U)
Lysinuric protein intolerance	Lys (U) markedly elevated	Cationic amino acids transporter SLC7A7	Arginine, ornithine moderate increase (U), orotic acid (U)
Fanconi syndrome	All amino acids elevated (U)	Defects in proximal renal tubule	

U, urine.

Table 5.
Laboratory findings in renal aminoacidurias.

Disorder of glycine metabolism			
Disorder name	**Amino acid involved**	**Enzyme/transport defect**	**Additional biomarkers**
Nonketotic hyperglycemia (NKH)	Gly (B, CSF) high	Mutations in Gly cleavage system	Increased CSF/plasma Gly ratio

B, blood; CSF, cerebrospinal fluid.

Table 6.
Laboratory findings in NKH.

1.7 Renal aminoacidurias

Renal aminoacidurias are disorders have inherited defects that affect renal tubular reabsorption process. Thus these disorders are characterized by abnormal urinary amino acids.

2. Newborn screening

Early diagnosis may prevent serious implications of inborn errors of metabolism, including amino acids disorders and significantly decrease morbidity and mortality. Newborn screening is a public health program that facilities early diagnosis by identifying neonates with potential treatable inborn errors of metabolism at the very early stages of their lives [13, 14]. This practice helps to manage the disease even for neonates that do not have evident symptoms in the first days of their lives. Amino acids analysis has always been an important part of the newborn screening. The first PKU screening bacterial inhibition assay was invented by Robert Guthrie in the early '60s [15]. Since that time, screening for IEMs is performed worldwide. In the United States, newborn screening is a state-mandated public health program ensuring that all newborns are screened for certain inherited conditions at birth. The panel of screening conditions varies from state to state. The advisory committee on heritable disorders in newborns and children advises the Secretary of Health and human services uniform screening panel, which currently consist of 34 core disorders and 26 secondary disorders. The recommended panel includes multiple amino acids related disorders (**Table** 7).

Recommended uniform screening panel (RUSP)	**Argininosuccinate aciduria, citrullinemia type I homocystinuria (cystathionine-β-synthase), maple syrup urine disease, phenylketonuria/ hyperphenylalaninemia, tyrosinemia I**
Additional non-RUSP conditions	Nonketotic hyperglycinemia (NKH), prolinemia, hyperammonemia/ornithinemia/ citrullinemia (HHH)
Secondary	Defects of biopterin cofactor biosynthesis and regeneration, citrullinemia II, hypermethioninemia, tyrosinemia II, tyrosinemia III

Table 7.
List of amino acids disorders that are recommended by the Secretary of the Department of Health and Human Services (HHS) as a part of state universal newborn screening (NBS) program effective July 2018.

3. Quantitative amino acids analysis

Quantitative amino acids analysis is an important tool for diagnosis of amino acids disorders and nutritional monitoring of individuals with already established diagnosis. Amino acids can be detected in most biological fluids, however, the most common fluids for inborn errors of metabolism diagnostics and monitoring are blood, plasma, and urine. In some cases, cerebrospinal fluid (CSF) amino acid levels are also diagnostic (**Table 5**). Although each disorder is biochemically characterized by abnormal levels of a single or a few amino acids, quantitative a non-screening analysis, and interpretation is not restricted to those metabolites and consist from a panel of nearly 40 amino acids and specific ratios. For example, along with plasma phenylalanine level, it is important also to assess plasma phenylalanine/tyrosine ratio that can be used to differentiate between PKU and non-PKU hyperphenylalaninemia [16].

3.1 Factors affecting amino acid analysis

The different chemical characteristics, a wide range of normal physiological levels [17–19], age groups variability and other factors detailed below represent a significant analytical challenge for the amino acid analysis. Diet is one of the significant factors that can highly affect amino acids levels [20, 21]. For example, meat and poultry consumption leads to increased excretion of β alanine and 1-methylhistidine. Thus blood collection intended for amino acids analysis is recommended after overnight fasting. Other factors such as urinary bacterial contamination can significantly alter urinary amino acids profile [22]. Some drugs interfere with amino acids metabolism [23] or cause signal artifacts. Valproic acid, for example, can cause an increase in plasma glycine. Anticoagulants used during sample collection also can contain interfering constituents [24]. For example, blood collection tubes containing sodium bisulfate in addition to heparin can yield a peak of S-sulfocysteine, falsely suggesting sulfite oxidase deficiency. Ethylenediaminetetraacetic acid (EDTA) additive in collection tube can produce ninhydrin-positive peaks, therefore lithium-heparin coated tubes are strongly preferred for the blood collection. An additional interfering factor to the amino acid analysis is a hemolysis as it may lead to the decrease of arginine with simultaneous increase of ornithine due to red blood cells arginase activity, and an increase in taurine that released from leukocytes and platelets. Serum is usually not a choice for the amino acids analysis, because blood needs to clot at room temperature during which asparagine is converted to aspartic acid and glutamine to glutamic acid.

For the urine analysis, a 24-h urine collection is preferred, alternatively, an overnight collection can be sufficient for the diagnostic purposes. In order to avoid artifacts, no preservatives are added to the urine sample.

Overall, during a prolong sample storage glutamine and asparagine decrease whereas glutamic and aspartic acids increase simultaneously. Additional markers of prolong storage are an increase of ethanolamine derived from phosphoethanolamine decomposition, increased tryptophan, GABA and taurine.

When cerebrospinal fluid is used for the analysis, it must be not contaminated with blood, as it leads to the nonspecific increase of multiple amino acids and can mask diagnostic findings.

Quantitative amino acids analysis implies in a variety of nonclinical fields such as biomedical research, bioengineering, food science, and agriculture. Multiple analytical methods have been developed over the years, however, some of these methods are not cost effective and labor intensive and thus are not applicable in clinical settings. The aim of next paragraphs is to describe the most common and widely used platforms in laboratory medicine field.

3.2 Ion exchange chromatography coupled with optical detection

In early 50s, diagnostic quantitative amino acid analysis became feasible with Moore and Stein publication on plasma amino acids separation with polystyrene resin column [25] and the subsequent automatization of the technique [26]. The principle, called ion exchange chromatography (IEC) with a post-column derivatization, for a long time remained a gold standard for the clinical amino acids analysis. Nowadays, despite the methodological advancements, the ion exchange chromatography using a lithium buffer system, followed the post-column derivatization with ninhydrin and UV detection is still widely used in clinical setting.

Standard sample preparation for IEC amino acid analysis involves deproteinization with 35% (w/v) sulfosalicylic acid (SSA) added to the biological fluid. It is recommended to use one volume of SSA to 10 volumes of plasma. A fixed amount of non-physiological amino acid as an internal standard is added to all samples. Commonly used internal standards are D-glucosaminic acid, S-2-aminoethyl-1-cysteine, norvaline, and norleucine, however, norleucine can interfere with argininosuccinic acid peak. After a short incubation, centrifugation and filtration, the sample is ready for the injection and separation.

In IEC, the separation is driven by the ionic interactions between the amino acid and functional ligands linked to the stationary phase of the column. The chromatographic column is filled with negatively charged resins. The sample is loaded on the column in low acidic pH and at this point, all amino acids bear a positive charge and strongly interact with the column. Manipulation with a lithium buffer composition during the run alters pH and salt composition, and as a result, there is a change in amino acid charge status (**Figure 5**). As the isoelectric point is reached amino acid is not charged anymore and has weak interactions with the charged column.

The complex separation of multiple amino acids is achieved based on ionic interactions strength. Amino acids with the weakest ionic interactions to the column start to elute first. After column elution, amino acids are mixed with a post-column reagent and are optically detected. The most common and well-established post-column derivatization is reaction with ninhydrin that produces a purple Ruhemann's chromophore (λ_{max} = 570 nm, **Figure 6**) for α-amino acids and yellow product with secondary amines (λ_{max} = 440 nm) for such as proline and hydroxyproline [27].

The absorbance intensity of the produced colorful analyte originated from every eluted amino acid is proportional to the amino acid's concentration in the examined biological fluid. Despite the fact that IEC amino acids technique is highly reproducible with a good linearity over a broad range, it suffers from a long run time for the full amino acids profile (about 150 min), and a lack of specificity as amino acids identification is based solely on retention time. Furthermore, co-elution of some amino acids on standard IEC method is

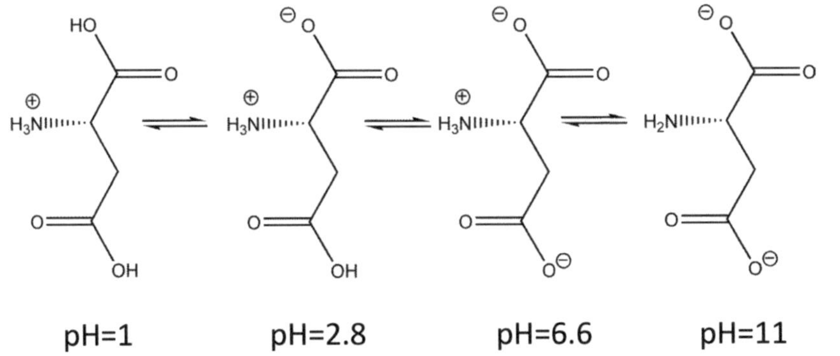

Figure 5.
Aspartic acid charge in different pH.

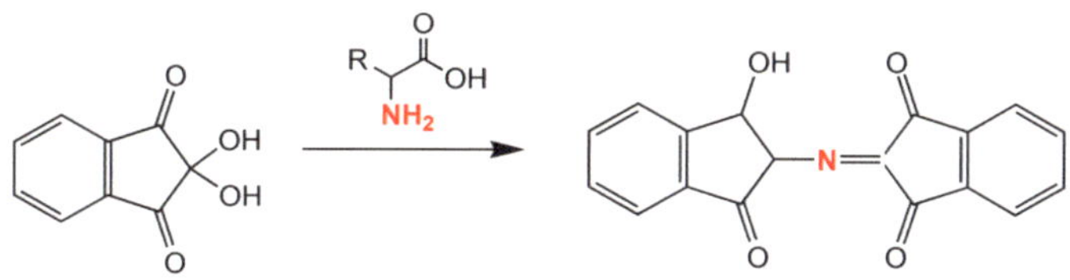

Figure 6.
Ninhydrin reaction with amino acid to produce Ruhemann's purple.

observed. For example, homocitrulline co-elutes with methionine and make challenging HHH syndrome. Moreover, allo-isoleucine, a diagnostic market for MSUD co-elutes with cystathionine. Argininosuccinic acid that accumulates in patients with argininosuccinate lyase deficiency has the same retention time as leucine. Additional drawback of the methodology is a limited stability of ninhydrin (recommended storage of the working solution ≤1 month) which adds up to the cost of the analysis.

3.3 RP-HPLC and RP-UPLC techniques

In recent years, reverse-phase high-performance liquid chromatography (RP-HPLC) and ultra-high performance liquid chromatography (UPLC) methods emerged as an alternative to the ion exchange chromatography. In RP-HPLC methods, the separation is based on hydrophobic interactions between the analyzed amino acid in the mobile phase and the immobilized hydrophobic ligands attached to the nonpolar column stationary phase. RP-HPLC offers a great resolution of very closely related molecules under a wide range of chromatographic conditions. For the optical detection, derivatization with o-phthalaldehyde (OPA) can be used as a pre-column or a post-column reaction. During the reaction, in the presence of thiol such as 2-mercaptoethanol, a stable fluorescent product is produced and can be detected by fluorimetry (excitation 340 nm and emission 410 nm) or UV (340 nm) [28, 29]. Although reproducible and automated [30], OPA derivatization method is not a good choice for proline/hydroxyproline and sulfur-containing amino acids detection. Alternative reagents for RP-HPLC with pre-column derivatization are phenylisothiocyanate (PITC, Pico-Tag commercialized by Waters) [31], dimethylamino-azobenzenesulfonyl-chloride (DABS-Cl) [32] and 9-fluorenylmethylchloroformate (FMOC-Cl) [33].

More advanced UPLC systems employ a small particle size (typically 1.7 μM) and a high pH range stable columns. These systems use less solvent and are operated in a high pressure which allows an excellent resolution achieved in a short time frame and thus potentially decreases turnaround time per sample. Narayan et al. analyzed 170 patient samples by pre-column 6-aminoquinolyl-*N*-hydroxysuccinimidyl carbamate (AQC) derivatization (**Figure 7**) followed by reverse phase UPLC [34] and compared amino acids data to the traditional amino acids analyzer operated through ion

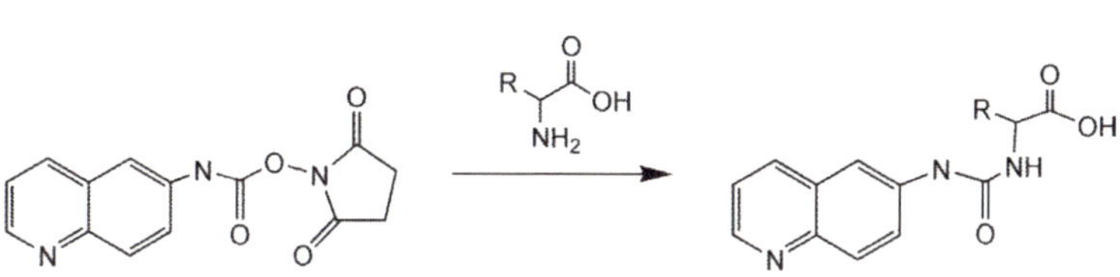

Figure 7.
AQC reaction with amino acid.

exchange chromatography method. The study found that UPLC method is comparable to the reference IEC and thus adaptable to the clinical laboratory.

Peake et al. developed a modified RP-UPLC method and achieved a better resolution for tyrosine, glycine, arginine and homocitrulline peaks [35]. The improved method also provides enhanced resolution to separate ornithine from mesocystathionine. There is a high clinical significance to accurate ornithine analysis as ornithine's levels are diagnostic for hyperornithinemia-hyperammonemia-homocitrullinemia (HHH) syndrome. The developed UPLC method has several advantages. Due to the short analysis time, it is feasible to include calibration prior to the analysis of urgent samples with a special turnaround times. Overall, RP-UPLC decreases turnaround time per sample, however, commercial kit components have a very limited shelf life and thus the method is not cost effective for clinical laboratories with a small samples volume.

Overall, ion exchange chromatography, RP-HPLC, and RP-UPLC techniques have a good reproducibility and a high sensitivity in the low picomole range, however, they all are carried out with optical detection. The main drawback of this type of detection is a lack of specificity as amino acids identification is solely based on the retention time. This can potentially cause to the false findings. For example, in a standard ion exchange chromatography method, ampicillin and amoxicillin co-elute with phenylalanine and it can be reported as falsely elevated.

3.4 Flow infusion tandem mass spectrometry (FIA-MS/MS)

More recently, developments and advancements in mass spectrometry field led to the inclusion of tandem mass spectrometry (MS/MS) as an alternative high throughput and specific technique for the amino acids analysis. It is also feasible to separate amino acids by liquid chromatography prior to the mass spectrometry analysis, however, it is time-consuming in clinical settings. Instead, tandem mass spectrometry scans are used for the high throughput, cost-effective amino acids analysis. It has to be noted, that FIA-MS/MS is a screening analysis that widely implemented through the newborn screening initiative.

For the newborn screening, blood samples are typically collected on filter paper and a defined size (typically 3 mm) disks are punched out of the paper and are extracted. The early assays required derivatization by butylation (**Figure 8**) in order to improve detection limits and minimize ion suppression effects in a complex biological matrix. Currently, to increase a throughput, some clinical laboratories skip on the derivatization step. Extracted and derivatized samples are directly introduced by injection to the mass spectrometer instrument with no chromatographic separation. Usually, 5–10 μl of a sample is injected into a flowing solvent at a very low (20–50 μl) flow rate. All screened amino acids (**Table 8**) are eluting at the same time whereas a typical run time is 1.5–2 min per sample. Every analyzed amino acid is assayed with the corresponding stable isotopic labeled standard.

The isotopic-labeled standards are closely related to the structure of the analyzed amino acids and have similar physicochemical properties to the target amino acids, but can be distinguished by mass spectrometry as they have a different

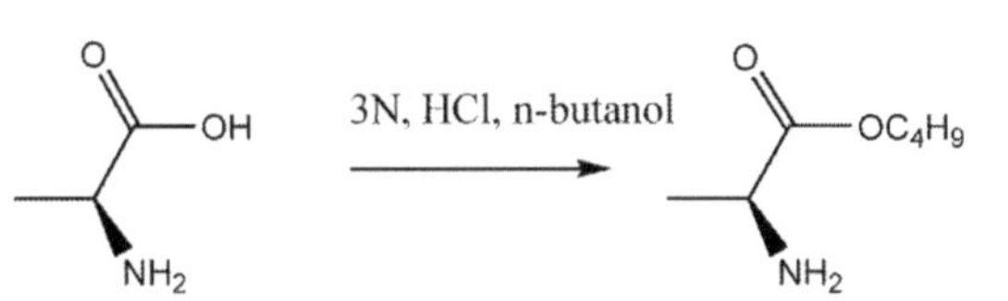

Figure 8.
Derivatization of alanine with n-butanol.

Target amino acid	*m/z*	Internal standard	*m/z*
Alanine	146.1	$^2H_4$ alanine	150.1
Arginine	231.2	$^{13}C,^2H_4$ arginine	236.2
Aspartic acid	246.2	$^2H_3$ aspartic acid	249.2
Citrulline	232.2	$^2H_2$ citrulline	234.2
Glutamic acid	260.2	$^2H_3$ glutamic acid	263.2
Glycine	132.1	$^{15}N,^{13}C$ glycine	134.1
Leucine/isoleucine	188.2	$^2h_3$ leucine	191.2
Methionine	206.2	$^2H_3$ methionine	209.2
Ornithine	189.2	$^2H_2$ ornithine	191.2
Phenylalanine	222.2	$^{13}C_6$ phenylalanine	228.2
Tyrosine	238.2	$^{13}C_6$ tyrosine	244.2
Valine	174.2	$^2H_8$ valine	182.2

Table 8.
Amino acids analyzed by FIA-MS/MS for the standard newborn screening panel and their stable isotopic labeled internal standards.

mass to charge ratio (*m/z*) (**Table 8**). They are added at a known quantity, and the response of each analyzed amino acid is normalized by the response of the matching internal standard. This type of normalization reduces a systematic error due to the poor recovery and decreases multiple matrix effects. The inclusion of internal standards also corrects a batch to batch variability due to the sample preparation and overall raises the accuracy and precision of the assay.

The tandem mass spectrometer has five basic components: the ion source where all molecules are a subject to the soft ionization, a mass analyzer that separates analytes based on their mass to charge ratio (Q1), a collision cell where molecular ions encounter an inert gas and undergo fragmentation (Q2), a second mass analyzer to separate fragments produced in the collision cell (Q3), and a detector. In collision cell, most of the screened butylated α-amino acids form a common and a very specific fragment of 102 Da (**Figure 9**). The tandem mass spectrometer then can be set to scan a constant mass difference of 102 Da and to produce a spectrum of the molecular ions derived from those amino acids that lost 102 Da in the collision cell (Q2) (**Figure 9**). Butylated amino acids with a basic side chains such as ornithine, citrulline loose ammonia and butyl formate in the

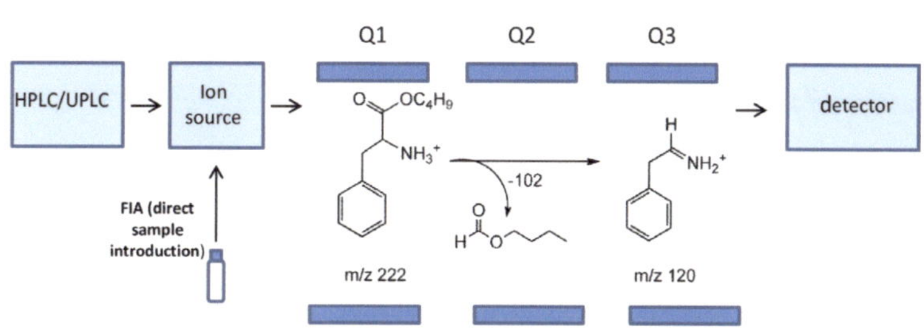

Figure 9.
Schematic presentation of tandem mass spectrometer. Phenylalanine (as butyl ester) looses 106 Da in the collision cell. When mass spectrometer operates in neutral loss scanning mode, it scans Q1 and Q3 in a synchronized manner. The mass difference of 102 Da (corresponds to a neutral fragment common to the most amino acids) passing through Q2 remains constant.

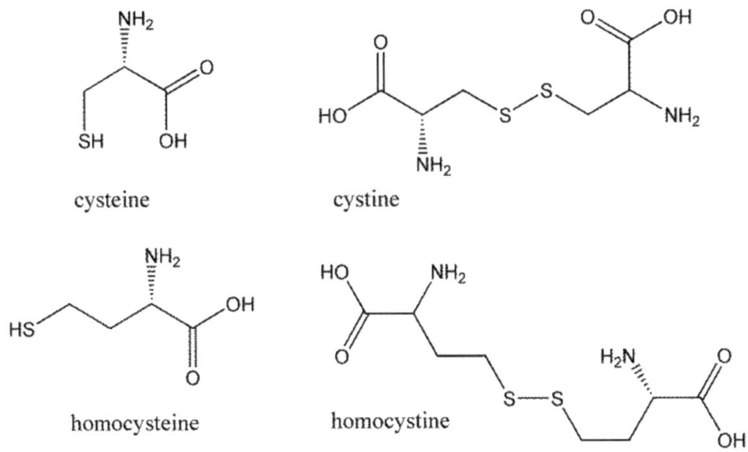

Figure 10.
Sulfur amino acids and their disulfides.

collision cell (*m/z* 119). For glycine and arginine, the most intensive signal corresponds to the loss of 56 and 161 Da, respectively. All these specific losses or transitions can be detected by different and highly specific tandem mass spectrometer's scans in the single analysis.

The main limitation of the FIA-MS/MS is inability to differentiate amino acids that share the same *m/z* such as leucine/isoleucine and hydroxyproline (butylated derivatives *m/z* 188), alanine/sarcosine (butylated derivatives *m/z* 146) and in a more extended profiles glutamine/lysine (butylated derivatives *m/z* 186), proline/asparagine (butylated derivatives *m/z* 172). Also, FIA-MS/MS is not applicable for cysteine and homocysteine analysis since these amino acids are not stable and react to form cystine and homocystine (**Figure 10**). During the ionization process, cystine and homocystine produce double charged molecules and it complicates the analysis.

Due to the high sensitivity and selectivity, there are more mass spectrometry-based techniques are available for the amino acids analysis, although because of extensive sample preparation or limited number of amino acids covered, these methods are not widely used in clinical laboratories. Gas chromatography mass spectrometry (GCMS) [36], capillary electrophoresis mass spectrometry (CEMS) [37], ion pairing (IP)-LC-MS/MS, HILIC-LC-mass spectrometry [38] and two column LC-MS/MS methods [39], ion pairing (IP)-LC-HRMS (TOF) [40] can be successfully applied for the physiological amino acids analysis although with some limitations.

4. Diagnosis of amino acids related disorders

The initial diagnosis of amino acids disorders is based on clinical presentation and biochemical findings such as abnormal levels of specific amino acids (**Tables 1–6**) or accumulation of the downstream metabolites in biological fluids, however, these characteristics are very heterogenic and often nonspecific. The most common clinical indications for the quantitative amino acid analysis in neonates and pediatric patients are coma, lethargy, seizures and vomiting, unexplained developmental delay and siblings with similar symptoms. Plasma amino acids analysis is also ordered as a conformational test to follow up abnormal newborn screening results. Hyperammonemia is characteristic to the most urea cycle disorders (**Table 3**) and therefore is another

DOI: http://dx.doi.org/10.5772/intechopen.84672

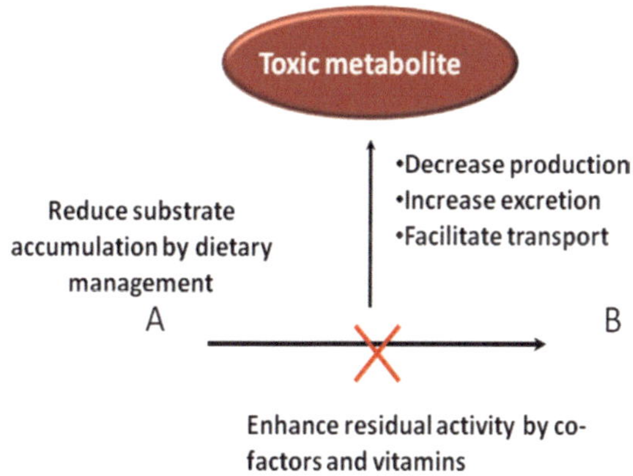

Figure 11.
Treatment strategies in amino acids disorders.

strong indication for plasma amino acids analysis. Additional general biochemical indicators of follow up quantitative amino acids analysis are ketosis (high blood and urine ketones), acidosis (blood pH below 7.35) and lactic acidemia (high lactate excretion), alkalosis (blood pH above 7.45), polyuria, polydipsia (extreme thirstiness), and dehydration. Amino acids analysis is also an important tool in the diagnosis of muscle and liver diseases, neurological disorders, renal failure, autism spectrum disorders and nutritional disturbances. Interpretation of amino acids profile is not just based on the abnormal level of a single amino acid, but rather involves pattern recognition, diagnostic ratios (**Tables 1** and **2**) and correlation to the patient's clinical history. It is recommended to confirm the diagnosis by molecular analysis or *in vitro* enzymatic assay (usually skin or tissue biopsy sample or blood cells).

5. Treatment options

Currently, there are numbers of available therapeutic approaches that aim in a substrate and downstream products restoration balance (**Figure 11**). One of the

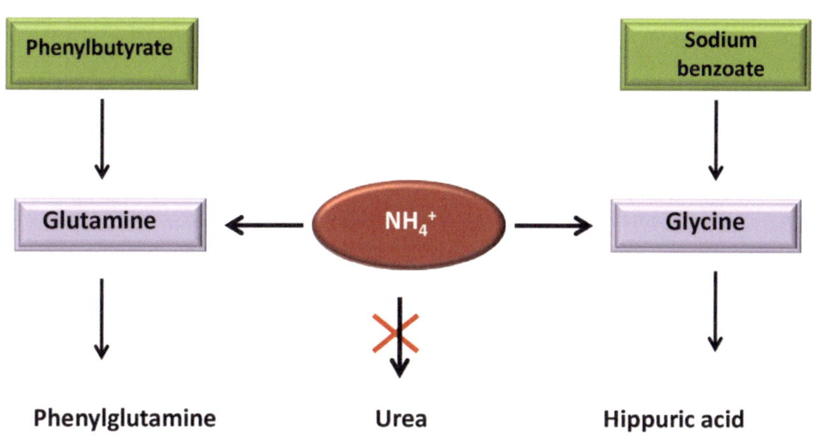

Figure 12.
Removal of toxic ammonia. In urea cycle disorders ammonia cannot be converted to urea, but alternatively can be converted to glutamine and glycine. Ammonia scavengers phenylbutyrate and sodium benzoate react with glutamine and glycine and consequently remove excess of ammonia. Phenylglutamine and hippurate are excreted in urine.

approaches in this direction is to reduce substrate accumulation by dietary restrictions. Nutritional therapies restrict offending amino acid or often total protein consumption through provision and monitoring of all essential components to meet dietary requirements. For example, special medical foods for PKU affected individuals have a very negligible amount of phenylalanine, but supplement the total protein required for the normal growth, development and nutritional status. Another example is MSUD nutritional management that restricts intake of the branch chain amino acids [41], but supplies the majority of the protein required in the standard diet.

Amino acids disorders are often manifested by the accumulation of toxic downstream metabolites. For example, urea acid disorders are characterized by life-threatening hyperammonemia (ammonia accumulation). Toxic metabolites removal treatment aims at reducing production or increasing excretion of these metabolites. To reduce hyperammonemia, sodium benzoate and phenylbutyrate are used to increased ammonia excretion (**Figure 12**) and to bypass the urea cycle metabolic block [42, 43]. Another example is an approach to reduce the production of succinylacetone, a neurological toxin that accumulates in tyrosinaemia I. Nitisinone (NTBC) treatment blocks a formation of fumarylacetoacetate and its subsequent conversion to the succinylacetone [44, 45].

If as a result of mutation, a specific enzyme still retains its residual activity, it can be stimulated by a co-factor or a co-factor precursor supplementation. This concept applies in treating tetrahydrobiopterin deficiency (**Figure 2**) [46, 47], remethylation defects (**Figure 4**) [48] and cystathionine beta-synthase deficiency (**Figure 4**) [10]. In some amino acids disorders, even partial metabolic block prevents from synthesizing an essential downstream metabolite to meet metabolic requirements. In these cases, essential product supplementation is required. For example, as a part of urea cycle disorders management, L-arginine and L-citrulline are administered [43]. This helps to reduce excessive protein catabolism, due to the low arginine levels.

6. Conclusion

In conclusion, amino acids disorders are a group of inborn errors of metabolism with highly variable clinical and biochemical presentations. Clinical manifestation often comprises severe neurological symptoms, growth and developmental delays. Most of the amino acids disorders related conditions are included in the newborn screening program to facilitate early diagnosis and early disease treatment. The analysis of physiological amino acids levels is a key tool in the diagnosis and clinical management of inborn errors of amino acids metabolism. A small subset of amino acids is analyzed in newborn screening by tandem mass spectrometry and leads to the detection of affected neonates even when they do not have a symptomatic disease manifestation. A more comprehensive, quantitative amino acids analysis covers analysis of nearly 40 amino acids. Prior to the analysis and results interpretation, pre-analytical variables such as a fasting status and medication treatments should be taken into account in order to avoid false reported findings. A most common sample preparation method for the quantitative amino acids analysis is acidification of specimen with a known small volume of concentrated acid, such as sulfosalicylic acid to precipitate proteins and large molecules, followed by centrifugation, leaving the water-soluble amino acids in the supernatant for the analysis. A variety of analytical methods have been developed over the past 60 years, and scientists have made significant achievements in the fields of derivatization, chromatography and mass spectrometry, however, the ion exchange chromatography method still remains the gold standard technique in the field. It is expected that more advanced techniques will be developed targeting important clinical laboratories requirements such as reduced samples pretreatment, linearity over the large concentration range for over the 40 amino acids, increased automation, high sensitivity, shorter run time and improved specificity. These methodological improvements will facilitate the diagnostic process and therapy monitoring for amino acids disorders. The field is also expending to the more exploratory platforms such as a whole-genome sequencing and untargeted metabolomics. Although these modalities have some restrictions in clinical settings [1, 49], they facilitate novel genes identification, novel biomarkers discovery and disease associations and thus strongly advancing the field [50, 51].

The major treatment goal for amino acids disorders is to normalize imbalance between the substrate and its downstream products and to avoid accumulation of the toxic substances. At the same time, nutritional management must meet basic dietary requirements for growth and normal development. Even though for many amino acids disorders current treatments do not offer a cure, they significantly improve the quality of life. It is expected over the upcoming years, that methodological advances will lead to a greater understanding of the IEM and in particular amino acids related disorders which will help further to improve disease outcomes.

Author details

Yana Sandlers
Cleveland State University, Cleveland, OH, USA

*Address all correspondence to: y.sandlers@csuohio.edu

References

[1] Sandlers Y. The future perspective: Metabolomics in laboratory medicine for inborn errors of metabolism. Translational Research. 2017. DOI: 10.1016/j.trsl.2017.06.005

[2] National Institutes of Health Consensus Development Panel. National Institutes of Health Consensus Development Conference Statement: Phenylketonuria: Screening and management, October 16-18, 2000. Pediatrics. 2001;**108**:972-982. Available from: http://www.ncbi.nlm.nih.gov/pubmed/11581453 [Accessed: 20 November 2018]

[3] Levy HL, Ghavami M. Maternal phenylketonuria: A metabolic teratogen. Teratology. 1996;**53**:176-184. DOI: 10.1002/(SICI)1096-9926(199603)53:3<176::AID-TERA5>3.0.CO;2-2

[4] Longo N. Disorders of biopterin metabolism. Journal of Inherited Metabolic Disease. 2009;**32**:333-342. DOI: 10.1007/s10545-009-1067-2

[5] De Braekeleer M, Larochelle J. Genetic epidemiology of hereditary tyrosinemia in Quebec and in Saguenay-Lac-St-Jean. American Journal of Human Genetics. 1990;**47**:302-307. Available from: http://www.ncbi.nlm.nih.gov/pubmed/2378355 [Accessed: 20 November 2018]

[6] Natt E, Kida K, Odievre M, Di Rocco M, Scherer G. Point mutations in the tyrosine aminotransferase gene in tyrosinemia type II. Proceedings of the National Academy of Sciences of the United States of America. 1992;**89**:9297-9301. DOI: 10.1073/PNAS.89.19.9297

[7] Tomoeda K, Awata H, Matsuura T, Matsuda I, Ploechl E, Milovac T, et al. Mutations in the 4-hydroxyphenylpyruvic acid dioxygenase gene are responsible for tyrosinemia type III and hawkinsinuria. Molecular Genetics and Metabolism. 2000;**71**:506-510. DOI: 10.1006/mgme.2000.3085

[8] Rüetschi U, Cerone R, Pérez-Cerda C, Schiaffino MC, Standing S, Ugarte M, et al. Mutations in the 4-hydroxyphenylpyruvate dioxygenase gene (HPD) in patients with tyrosinemia type III. Human Genetics. 2000;**106**:654-662. Available from: http://www.ncbi.nlm.nih.gov/pubmed/10942115 [Accessed: 20 November 2018]

[9] Blackburn PR, Gass JM, Vairo FPE, Farnham KM, Atwal HK, Macklin S, et al. Maple syrup urine disease: Mechanisms and management. The Application of Clinical Genetics. 2017;**10**:57-66. DOI: 10.2147/TACG.S125962

[10] Morris AAM, Kožich V, Santra S, Andria G, Ben-Omran TIM, Chakrapani AB, et al. Guidelines for the diagnosis and management of cystathionine beta-synthase deficiency. Journal of Inherited Metabolic Disease. 2017;**40**:49-74. DOI: 10.1007/s10545-016-9979-0

[11] Sloan JL, Carrillo N, Adams D, Venditti CP. Disorders of Intracellular Cobalamin Metabolism. Seattle: University of Washington; 1993. Available from: http://www.ncbi.nlm.nih.gov/pubmed/20301503 [Accessed: 26 December 2018]

[12] Refsum H, Smith AD, Ueland PM, Nexo E, Clarke R, McPartlin J, et al. Facts and recommendations about total homocysteine determinations: An expert opinion. Clinical Chemistry. 2004;**50**:3-32. DOI: 10.1373/clinchem.2003.021634

[13] Urv TK, Parisi MA. Newborn screening: Beyond the spot. Advances in Experimental Medicine and Biology. 2017;**1031**:323-346. DOI: 10.1007/978-3-319-67144-4_19

[14] Berry SA. Newborn screening. Clinics in Perinatology. 2015;**42**:441-453. DOI: 10.1016/j.clp.2015.03.002

[15] Blumenfeld CM, Wallace MJ, Anderson R. Phenylketonuria—The Guthrie screening test—A method of quantitation, observations on reliability and suggestions for improvement. California Medicine. 1966;**105**:429-434. Available from: http://www.ncbi.nlm.nih.gov/pubmed/18730038 [Accessed: 24 November 2018]

[16] Eastman JW, Sherwin JE, Wong R, Liao CL, Currier RJ, Lorey F, et al. Use of the phenylalanine: Tyrosine ratio to test newborns for phenylketonuria in a large public health screening programme. n.d. Journal of Medical Screening. 2000;**7**(3)131-135. Available from: www.jmedscreen.com [Accessed: 28 November 2018]

[17] Yamamoto H, Kondo K, Tanaka T, Muramatsu T, Yoshida H, Imaizumi A, et al. Reference intervals for plasma-free amino acid in a Japanese population. n.d. Annals of Clinical Biochemistry. 2016;**53**(3)357-364. DOI: 10.1177/0004563215583360

[18] Haschke-Becher E, Kainz A, Bachmann C. Reference values of amino acids and of common clinical chemistry in plasma of healthy infants aged 1 and 4 months. Journal of Inherited Metabolic Disease. 2016;**39**:25-37. DOI: 10.1007/s10545-015-9870-4

[19] Lepage N, McDonald N, Dallaire L, Lambert M. Age-specific distribution of plasma amino acid concentrations in a healthy pediatric population. Clinical Chemistry. 1997;**43**:2397-2402

[20] Conley TB, Apolzan JW, Leidy HJ, Greaves KA, Lim E, Campbell WW. Effect of food form on postprandial plasma amino acid concentrations in older adults. The British Journal of Nutrition. 2011;**106**:203-207. DOI: 10.1017/S0007114511000419

[21] Ottosson F, Ericson U, Almgren P, Nilsson J, Magnusson M, Fernandez C, et al. Postprandial levels of branch chained and aromatic amino acids associate with fasting glycaemia. Journal of Amino Acids. 2016;**2016**:1-9. DOI: 10.1155/2016/8576730

[22] Vidler J, Wilcken B. Prevalence of unsuspected urinary bacterial contamination: Effects of screening tests for detection of inborn errors of metabolism. Clinica Chimica Acta. 1978;**82**:173-178. Available from: http://www.ncbi.nlm.nih.gov/pubmed/618679 [Accessed: 28 November 2018]

[23] Edwards MA, Grant S, Green A. A practical approach to the investigation of amino acid disorders. 1988. Available from: https://journals.sagepub.com/doi/pdf/10.1177/000456328802500202 [Accessed: 28 November 2018]

[24] Chuang CK, Lin SP, Lin YT, Huang FY. Effects of anticoagulants in amino acid analysis: Comparisons of heparin, EDTA, and sodium citrate in vacutainer tubes for plasma preparation. Clinical Chemistry. 1998;**44**:1052-1056. Available from: http://www.ncbi.nlm.nih.gov/pubmed/9590384 [Accessed: 28 November 2018]

[25] Moore S, Spackman DH, Stein WH. Chromatography of amino acids on sulfonated polystyrene resins. An improved system. Analytical Chemistry. 1958;**30**:1185-1190. DOI: 10.1021/ac60139a005

[26] Moore S, Spackman DH, Stein WH. Automatic recording apparatus for use in the chromatography of amino acids. Federation Proceedings. 1958;**17**:1107-1115. Available from: http://www.ncbi.nlm.nih.gov/pubmed/13619781 [Accessed: 28 November 2018]

[27] Friedman M. Applications of the ninhydrin reaction for analysis of amino acids, peptides, and proteins to

agricultural and biomedical sciences. 2004. p. 385-406. DOI: 10.1021/JF030490P

[28] Cunico RL, Schlabach T. Comparison of ninhydrin and o-phthalaldehyde post-column detection techniques for high-performance liquid chromatography of free amino acids. Journal of Chromatography. A. 1983;**266**:461-470. DOI: 10.1016/S0021-9673(01)90917-3

[29] Dorresteijn RC, Berwald LG, Zomer G, de Gooijer CD, Wieten G, Beuvery EC. Determination of amino acids using o-phthalaldehyde-2-mercaptoethanol derivatization effect of reaction conditions. Journal of Chromatography A. 1996;**724**:159-167. DOI: 10.1016/0021-9673(95)00927-2

[30] Frank MP, Powers RW. Simple and rapid quantitative high-performance liquid chromatographic analysis of plasma amino acids. Journal of Chromatography. B, Analytical Technologies in the Biomedical and Life Sciences. 2007;**852**:646-649. DOI: 10.1016/j.jchromb.2007.01.002

[31] Sherwood RA. Amino acid measurement in body fluids using PITC derivatives. In: Amino Acid Analysis Protocols. New Jersey: Humana Press; 2000. pp. 169-175. DOI: 10.1385/1-59259-047-0:169

[32] Schneider H-J. Amino acid analysis using DABS-CL. Chromatographia. 1989;**28**:45-48. DOI: 10.1007/BF02290382

[33] Bütikofer U, Fuchs D, Bosset JO, Gmür W. Automated HPLC-amino acid determination of protein hydrolysates by precolumn derivatization with OPA and FMOC and comparison with classical ion exchange chromatography. Chromatographia. 1991;**31**:441-447. DOI: 10.1007/BF02262386

[34] Narayan SB, Ditewig-Meyers G, Graham KS, Scott R, Bennett MJ. Measurement of plasma amino acids by Ultraperformance® Liquid Chromatography. Clinical Chemistry and Laboratory Medicine. 2011;**49**: 1177-1185. DOI: 10.1515/CCLM.2011.200

[35] Peake RWA, Law T, Hoover PN, Gaewsky L, Shkreta A, Kellogg MD. Improved separation and analysis of plasma amino acids by modification of the MassTrak™ AAA Solution Ultraperformance® liquid chromatography method. Clinica Chimica Acta. 2013;**423**:75-82. DOI: 10.1016/J.CCA.2013.03.036

[36] Kaspar H, Dettmer K, Chan Q, Daniels S, Nimkar S, Daviglus ML, et al. Urinary amino acid analysis: A comparison of iTRAQ®-LC-MS/MS, GC-MS, and amino acid analyzer. Journal of Chromatography B. 2009;**877**:1838-1846. DOI: 10.1016/j.jchromb.2009.05.019

[37] Hirayama A, Soga T. Amino acid analysis by capillary electrophoresis-mass spectrometry. Methods in Molecular Biology. 2012;**828**:77-82. DOI: 10.1007/978-1-61779-445-2_8

[38] Prinsen HCMT, Schiebergen-Bronkhorst BGM, Roeleveld MW, Jans JJM, de Sain-van der Velden MGM, Visser G, et al. Rapid quantification of underivatized amino acids in plasma by hydrophilic interaction liquid chromatography (HILIC) coupled with tandem mass-spectrometry. Journal of Inherited Metabolic Disease. 2016;**39**:651-660. DOI: 10.1007/s10545-016-9935-z

[39] Le A, Ng A, Kwan T, Cusmano-Ozog K, Cowan TM. A rapid, sensitive method for quantitative analysis of underivatized amino acids by liquid chromatography-tandem mass spectrometry (LC-MS/MS). Journal of Chromatography B. 2014;**944**:166-174. DOI: 10.1016/j.jchromb.2013.11.017

[40] Armstrong M, Jonscher K, Reisdorph NA, Reisdorph NA. Analysis of 25 underivatized amino acids in human plasma using ion-pairing reversed-phase liquid chromatography/time-of-flight mass spectrometry. Rapid Communications in Mass Spectrometry. 2007;**21**:2717-2726. DOI: 10.1002/rcm.3124

[41] Strauss KA, Wardley B, Robinson D, Hendrickson C, Rider NL, Puffenberger EG, et al. Classical maple syrup urine disease and brain development: Principles of management and formula design. Molecular Genetics and Metabolism. 2010;**99**:333-345. DOI: 10.1016/J.YMGME.2009.12.007

[42] Peña-Quintana L, Llarena M, Reyes-Suárez D, Aldámiz-Echevarria L. Profile of sodium phenylbutyrate granules for the treatment of urea-cycle disorders: Patient perspectives. Patient Preference and Adherence. 2017;**11**:1489-1496. DOI: 10.2147/PPA.S136754

[43] Häberle J, Boddaert N, Burlina A, Chakrapani A, Dixon M, Huemer M, et al. Suggested guidelines for the diagnosis and management of urea cycle disorders. Orphanet Journal of Rare Diseases. 2012;7:32. DOI: 10.1186/1750-1172-7-32

[44] Lindstedt S, Holme E, Lock EA, Hjalmarson O, Strandvik B. Treatment of hereditary tyrosinaemia type I by inhibition of 4-hydroxyphenylpyruvate dioxygenase. Lancet. 1992;**340**:813-817. Available from: http://www.ncbi.nlm.nih.gov/pubmed/1383656 [Accessed: 16 January 2019]

[45] Masurel-Paulet A, Poggi-Bach J, Rolland M-O, Bernard O, Guffon N, Dobbelaere D, et al. NTBC treatment in tyrosinaemia type I: Long-term outcome in French patients. Journal of Inherited Metabolic Disease. 2008;**31**:81-87. DOI: 10.1007/s10545-008-0793-1

[46] van Vliet D, Anjema K, Jahja R, de Groot MJ, Liemburg GB, Heiner-Fokkema MR, et al. BH4 treatment in BH4-responsive PKU patients: Preliminary data on blood prolactin concentrations suggest increased cerebral dopamine concentrations. Molecular Genetics and Metabolism. 2015;**114**:29-33. DOI: 10.1016/j.ymgme.2014.11.009

[47] Christ SE, Moffitt AJ, Peck D, White DA. The effects of tetrahydrobiopterin (BH4) treatment on brain function in individuals with phenylketonuria. NeuroImage: Clinical. 2013;**3**:539-547. DOI: 10.1016/j.nicl.2013.08.012

[48] Ogier de Baulny H, Gérard M, Saudubray JM, Zittoun J. Remethylation defects: Guidelines for clinical diagnosis and treatment. European Journal of Pediatrics. 1998;**157**(Suppl 2):S77-S83. Available from: http://www.ncbi.nlm.nih.gov/pubmed/9587031 [Accessed: 17 January 2019]

[49] Vernon HJ. Inborn Errors of Metabolism Advances in Diagnosis and Therapy. 2015;**169**(8):778-782. DOI: 10.1001/jamapediatrics.2015.0754

[50] Yubero D, Brandi N, Ormazabal A, Garcia-Cazorla À, Pérez-Dueñas B, Campistol J, et al. Targeted next generation sequencing in patients with inborn errors of metabolism. PLoS ONE. 2016;**11**:e0156359. DOI: 10.1371/journal.pone.0156359

[51] Miller MJ, Kennedy AD, Eckhart AD, Burrage LC, Wulff JE, Miller LAD, et al. Untargeted metabolomic analysis for the clinical screening of inborn errors of metabolism. Journal of Inherited Metabolic Disease. 2015;**38**:1029-1039. DOI: 10.1007/s10545-015-9843-7

Chapter 9

Resource-Based View of Laboratory Management: Tissue Bank ATMP Production as a Model

Wojciech Smętek, Jacek Węgrzyk, Agnieszka Klama-Baryła, Wojciech Łabuś, Małgorzata Kraut, Michał Szapski, Mariusz Nowak and Diana Kitala

Abstract

Modern health care organizations, e.g., tissue banks, require a resource-based view (RBV) for an efficient stimulation of innovation, productivity, and performance, especially in the context of laboratory management and new product development. High quality advanced therapy medicinal products (ATMPs) are expected to bring important health benefits; therefore, their production has to be performed in accordance with good manufacturing practice (GMP). Although there are no precisely defined criteria for quality control/evaluation methods of obtained ATMPs, all aspects of pharmaceutical quality of ATMPs' development, manufacturing, distribution, inspection, and review processes ought to be strictly fulfilled. Explicit performance management and production regimes in accordance with pharmacopeia and RBV philosophy have been proposed in this chapter.

Keywords: resource-based view, advanced therapy medicinal products, tissue bank, good manufacturing practice, lean management

1. The change of paradigm

An innovative way of hospital management appeared in the 1990s. Methods originating from the industry and known as lean management and Theory of Constraints (TOC) have proven surprisingly adequate in health care environment. These helped to overcome the problems of many hospitals facing severe issues and as a result requiring turnaround.

Effective management of the organization has to be holistic in nature and focus exclusively on the point of view of client/patient instead of unit managers. The effectiveness and process synchronization have to be also looked upon from the same perspective. Management approach aiming at local optimizations traps the managers ultimately leading to underperformance of the organization as a whole, witch hunt and frustration from the very bottom to the top of the organizational ladder. Moreover, most of the employees including top managers are not fully aware of the structure of processes in their company. These processes do not fit the most commonly applied management practices. In their nature, these are horizontal and not vertical like the structures of most of the organizations [1].

IntechOpen

The term “lean management” has been first formulated by the American researcher John Krafcik who has been investigating the reasons behind competitive advantage of Japanese automobile industry over its American counterpart. Krafcik has been particularly impressed by the Toyota Production System (TPS), which has been developing in Japan from mid-1940s.

Essentially, lean system could be defined as an elimination of waste and respect to all of the employees organization-wide. The goal of the organization is to deliver added value to the patient/client/payer. Added value is generated when any activity that delivers tangible results in the process is performed correctly the first time and for which the patient/client/payer is ready to pay. All the rest is regarded as waste. There are seven types of waste defined by the TPS originator—Taiichi Ohno. Detailed description is outside the scope of this book, but an interested reader will easily find it outlined in many available resources [2].

The set of tools and techniques applied in lean is usually presented as the so-called Toyota House or Lean House. The fundaments of the house comprise 5S—the system of creating and maintaining a neat and clean work environment, continuous improvement, standard work, balanced distribution of the job, and visual controls. All of these elements are crucial; nevertheless, 5S and standardized work require closer explanation.

The majority of organizations do not realize how much waste is hidden in their current processes. There is an abundance of unnecessary tools, equipment, excess inventory, documentation, etc., in any hospital. All these obscure the real problems. Any attempt to optimize the processes without launching 5S at the very beginning is doomed to fail as the main issues will stay hidden behind the visible surface.

Standard work in lean is different from what is usually meant by the notion. Standardization is the best presently known way of performing a certain task. If the employees develop a better, safer way, the standard work can be changed almost immediately. This is different from the traditional approach relying on rigid rules, which are almost impossible to change. Actually, lean is a scientific method based on continuous experiments, which make it extremely flexible, and ensures involvement of the employees on all the levels.

Two walls of the Toyota House can be distinguished. The first one is continuous flow, which concentrates on elimination of delays as well as right care in the right time in the right place. The second one is a quality approach, focusing on error detection, preventing errors at the very source, and involving the employees in problem-solving. The most important rule is to ensure that no witch hunt takes place. Instead, employees providing information about the existing and possible issues and implementations of related solutions should be rewarded.

Finally, the roof comprises the ultimate goals: health, safety, staff morale, and patient satisfaction.

Unfortunately, the term “lean” has been misinterpreted, mainly in the USA, and used in the form of practices involving layoffs and increased work burden. As a result, it has lead to pathologies that are in no way connected to the Japanese ideas. The authors would like to stress that the aforementioned mismanagement has nothing in common with genuine lean, which is based directly on TPS.

Theory of Constraints emerged at the beginning of the 1980s, when the Israeli physicist, Dr. Eliyahu Goldratt who became interested in the production planning processes, published his first book on the subject in the form of a bestselling business novel *The Goal* [3]. Goldratt claimed that every organization has its own main constraint, either internal or external, and proposed a set of five steps called the Process of Ongoing Improvement (POOGI) to handle these issues accordingly. The aforementioned steps are:

- Identify the constraint,
- Exploit the constraint—ensure that the constraint is operational all the available time,
- Subordinate all the resources within the organization to the requirements of the constraint,
- Elevate the constraint, i.e., widen it, and
- If the constraint has been broken, identify the new one and start the process all over again, but be aware of the inertia.

Over the following years, Goldratt developed the applications of the TOC in various areas like project management, sales, marketing, and people management. In the 1990s, he introduced the TOC thinking processes—a set of tools aimed at enhancing the analysis of the problems and building solid, logical, time-proven solutions. Today, TOC has found its applications in many different areas both in manufacturing and in services like logistics, strategy, health care, education, and resocialization [4].

It has to be stressed that both lean and TOC—if they are to be implemented the proper way—have to be a company-wide effort closely supervised and led by the top management. Both clinical and nonclinical departments must be taken into account. Obviously, the positive cultural change of the organization will follow as an ultimate result of the implementation. The task is by no means easy, but the payoffs are well worth it. Potential benefits include 20–80% of productivity improvement, reduction of unnecessary length of stay, and significant reduction in inventory and space, and savings counted in millions of local currency are not uncommon if the methods are applied properly and given sufficient time [5]. One has to remember that both methods advocate continuous and growing improvement and are therefore not a one-time initiative.

Neither lean nor TOC is meant to replace the golden medical standards or the standards of pharmaceutical production like GMP. They are rather complementary tools and techniques, paving the way for the existing standards to rise to new levels.

Dr. Stanislaw Sakiel Burn Treatment Centre (CLO) based in Siemianowice Śląskie, Poland, is a top hospital specializing primarily in the treatment of burns and chronic wounds. CLO has been a pioneer in application of TOC and lean in hospital management since early 2016. According to the knowledge of the authors, it is the only hospital in Poland where the system not only is still functioning but also delivers constantly better results. The Bank of Tissues has been among the first units where the approach has been applied. The project was initiated in 2016. Despite the fact that GMP was already in place, it soon became apparent that there is still a huge room for improvement. The decision to implement 5S quickly delivered results. Process mapping followed, together with visible changes in the flow in local production site. Today, the flow is faster, job is better organized, and the breaks are sporadic instead of being a norm. Improved communication with the operating theater resulted in more on-time deliveries and less unnecessary additional transports. This in turn contributed to shorter surgical procedures and better overall conditions for patients and medical personnel alike.

2. Resource-based view (RBV) approach in health care organizations and tissue bank laboratory management

2.1 Resource-based view framework for health care organizations

The resource-based view (RBV) is a theoretical framework used to study and explain the competitive behavior of organizations [6, 7] that emerged in 1980s and 1990s, after the major works published by Barney, J. *Firm Resources and Sustained Competitive Advantage*, Prahalad and Hamel *The Core Competence of the Corporation*, and Wernerfelt, B. *The Resource-Based View of the Firm*. Since then, RBV approach has been widely used and appreciated in the management of private organizations. However, it may also bring a promising framework to implement in the context of large-scale quality improvement within the public health care entities. Although these theories may be novel for the health management field, including models of strategic management originally developed for private sector firms, their application within publicly funded settings may be problematic or require customization, but nevertheless provide interesting insights [8]. It may provide a sustainable solution to manage quality driven, highly complex, and environmentally turbulenced settings such as tissue bank laboratories. RBV approach may be the answer to maximize growth, quality, and performance in a way to cope with modern, rapid technological changes and competitive medical environments.

The tissue banking sector quality improvement efforts take place in interorganizational networks rather than in a large, vertically integrated organization. Therefore, tissue bank managers encounter significant difficulties in understanding high organizational complexity that cannot be reduced to simple cause-effect relations or one variable [8]. The challenge is not only to be in compliance with applicable codes and quality regulations of good manufacturing practice (GMP), state regulatory agency, or pharmaceutical regulatory requirements, but also at the same time to optimize your business operations and standardize workflows in processing laboratories. Navigating through this complicated regulatory milieu and at the same ensuring efficient manufacturing process along with sensible business cost balance may require a RBV theoretical framework. Resource-based view and regulatory compliance are not mutually exclusive but should be combined to deliver excellent performance improvements and complex approach to stimulate organization's competitiveness, productivity, and innovation.

2.2 RBV model

The RBV resource approach assumes that the success of the organization lies within the organization itself, or to be exact, in its valuable, intangible, and not perfectly imitable resources (VRIO condition) allowing it to achieve a sustainable competitive advantage [9]. Resource-based view (RBV) analyzes and interprets resources of the organizations to understand how organizations achieve sustainable competitive advantage. The RBV focuses on the concept of difficult-to-imitate attributes of the firm as sources of superior performance and competitive advantage [10, 11]. It takes an 'inside-out' view of firm-specific perspective on why organizations succeed or fail in the market place [12]. According to this approach, resources are given the superior role in achieving higher organizational performance. The RBV model presents some critical assumptions regarding resources, which are about to be met, in order to obtain efficient manufacturing process in line with regulatory compliance. The first stage of the RBV model categorizes resources within the organization into two basic types: tangible and intangible resources. The literature categorizes

Types of Tissue Bank Tangible and Intangible Resources and Capabilities	
Tangible resources and capabilites	**Examples**
Financial	• Ability to generate internal funds • Ability to raise external capital (i.a., EU funding)
Physical	• Location of the tissue bank • Location of offices and laboratories • Location of machines, equipment, and systems • Access to distibution channels (hospitals) • Access to starting materials (hospitals and forensic medicine labs)
Technological	• Possession of patents, trademarks, and trade secrets, • R&D labs
Organizational	• Quality, planning, and command and control systems • Integrated lab management information systems
Intagible resources and capabilities	**Examples**
Human	• Managerial and operational talents (employees) • Organizational culture
Innovation	• R&D capabilities to innovate new grafts, products, medical procedures, and services • Capabilities for organizational innovation and change
Reputational	• Perception of product quality and reliability among medical staff • Reliability among patients • Succesful product positioning among hospitals and medical staff • Domestic and international recognizability of the tissue bank • Reputation as a good employer • Reputation as an innovative unit
Adapted from P. M. Madhani, 2010, Resource Based View (RBV) of Competitive Advantage: An Overwiev	

Table 1.
Examples of tissue bank tangible and intangible resources and capabilities.

resources also into three main categories: tangible resources, intangible resources, and capabilities [13]. Capabilities refer to the skills organization possesses to coordinate the resources (i.a., teamwork, organizational culture, and trust); however, they are not clearly owned and difficult to measure. For the purpose of the RBV model in this paper, the author included capabilities into tangible and intangible resource category (**Table 1**).

Tangible resources belong to the organization and can be divided into fixed assets (i.e., tissue bank facilities, land, laboratory machinery, clean room systems, and furniture) and current assets including, i.a., capital equipment, laboratory equipment, inventory, and financial means. Such assets are also referred to as property-based resources [6]. Other divisions of tangible resources include also categories of physical and financial assets [14].

Intangible resources, also referred to as knowledge-based resources [15], consist of intellectual property and include, i.a., operational knowledge of the laboratory employees, process knowledge, procedures, accumulated experience, patents, tissue bank brand recognition, community relationship, trademark, and legal agreements. They can be further divided into infinite and finite intangible assets (**Figure 1**).

Tangible and intangible resources can be further grouped into various subcategories. One of them was first presented by Barney [13] and included physical resources (physical, technological, plant, and equipment), human capital resources (training, experience, and insights), and organizational capital resources (formal structure). An example of broader classification of resources and capabilities of a tissue bank can be found in **Table 1**.

According to RBV, not all resources of the firm will be strategic and, hence, sources of competitive advantage. Competitive advantage occurs only when there is a situation of resource heterogeneity and immobility [16]. The resource heterogeneity implies that organizations have varying capabilities and possess unique bunch of resources allowing them to design different strategies to obtain competitive advantage. Resource immobility may be understood as inability of competing firms to obtain resources from other firms. If the resource is not perfectly mobile

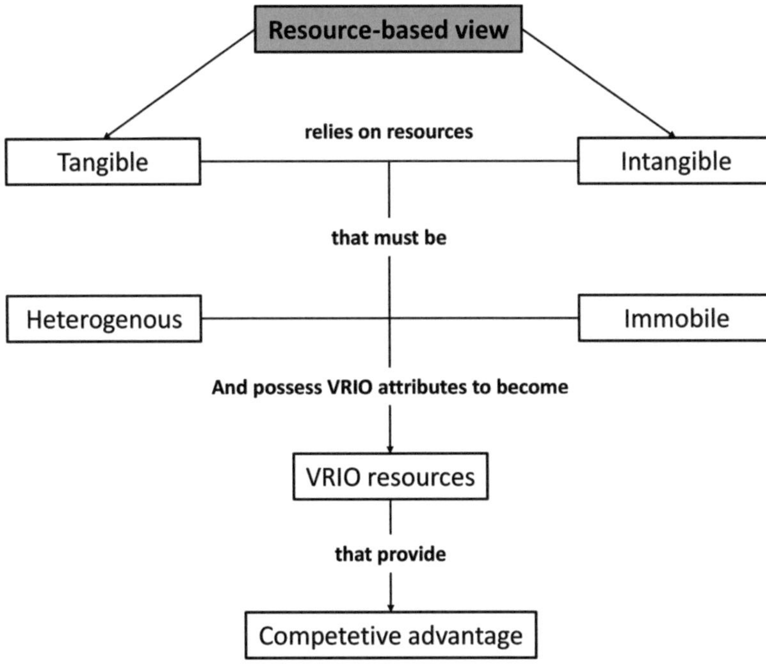

Figure 1.
RBV model.

(i.e., the resource is not free to move between firms, or if a firm without a resource faces a considerable cost burden in developing, acquiring, or using it that a firm already using it does not), then the resource is likely to be a source of sustained competitive advantage [16].

In order for the resource to provide sustainable and desirable performance, VRIO framework criteria must be fulfilled. The original tool VRIN was described by Barney [10] in his work *Firm Resources and Sustained Competitive Advantage* and was later improved by Barney to VRIO in *Looking Inside for Competitive Advantage* (1995).

1. **Valuable (V):** a resource is considered valuable when it provides strategic value to organization, enabling it to exploit opportunities or defend against threats. The resource that is unable to meet this criterion leads to competitive disadvantage.

2. **Rare (R):** rare resources, hard to obtain by competitors, grant temporary competitive advantage. The resource must be rare enough to design and execute unique business strategy in comparison with other organizations on the market.

3. **Imperfect imitability (I):** the resource leading to competitive advantage must be costly to imitate for other organizations that do not have it. The resource may also be hard to imitate not only because of financial reasons but also because of difficulties in acquiring it, its complexity, causal ambiguity, or even historical conditions.

4. **Non-Substitutability (N):** it is impossible for a competitor to substitute or replace a certain resource by other or alternative resource/**Organized to Capture Value (O):** it is the task of the organization to be able to organize its management systems, processes, and procedures in a way to fully realize the

Tissue Bank VRIO resource analysis example			
Patent: Production of Acellural Dermal Matrix (ADM) inhabited by in vitro culture cells			
Valuable?	**Rare?**	**Costly to imitate?**	**Is the organization organized to exploit it?**
Yes	Yes	Yes	Yes
The resource adds value to the organization by allowing it to perform and refund innovative and efficient ADM skin graft procedures.	The patent is controlled by the tissue bank and protected by patent regulations.	The research process of developing the production method is costly and time-consuming, and it requires specialized knowledge, equipment and workforce.	The tissue bank underwent general redecoration of the laboratories and equipment allowing it to produce ADM on a large scale. The organization possesses appropriate quality, production and management systems, and procedures to prepare and distribute the grafts in compliance with GMP and state regulatory agency requirements. The tissue bank employs qualified and experienced staff able to run the production process.
Result: sustained competitive advantage			

Table 2.
Examples of VRIO analysis: ADM patent.

potential of its unique resources. Therefore, the resource itself does not confer any value to the company; it is crucial to organize the company allowing it to achieve not only temporary but also sustained competitive advantage.

The VRIO framework may therefore be a crucial strategic analysis tool uncovering resources and capabilities that give organizations a long-term competitive advantage. An example of VRIO analysis for tissue bank can be found in **Table 2**.

2.3 Summary

Internal resources and capabilities determine strategic choices made by firms while competing in their external business environment [15]. As resource requirements in health care organizations must be carefully managed due to cost constraints, RBV approach along with VRIO resource analysis adaptation to tissue bank not-for-profit settings seems intuitively sound. Resource-based view can ensure sensible balance between complying quality regulations that govern tissue banks and optimizing processing practices. Examining and identifying unique (valuable, rare, and costly to imitate) organizational resources through RBV and VRIO framework may deliver simple but excellent performance improvements. Finding out if the entity is organized to exploit the resources and protecting them may lead to better financial results as control of superior and unique resources is necessary to establish a cost advantage, which leads to profitability [14]. Finally, constant review of VRIO resources and capabilities enables organizations to establish clear-cut sustainable competitive advantage and growth.

3. Activity and documentation of a tissue bank

The activity of a tissue bank may be limited to the preparation of biostatic (radiation sterilized) and biovital (living) grafts using documented and proven procedures. In this regard, a tissue bank operates under a Ministry of Health License and must demonstrate the implementation of a quality assurance system that meets the requirements of the Law of July 1, 2005, on the collection, storage, and transplantation of cells, tissues, and organs as amended (*Journal of Laws 2017 item 1000*). The regulation contains the basic documents that a tissue bank quality system must include; however, some of the mentioned documents seem to be obsolete. The basic documents used in a tissue bank are standard operating procedures (SOPs). SOPs must describe the requirements regarding primarily the processes and activities related to the collection of cells or tissues, their acceptance, processing, storage, distribution, transport, as well as monitoring the condition of cells or tissues on their way from the donor to the recipient. Standard operating procedures should specify the medical devices and materials that have direct contact with cells or tissues. The criteria for the qualification and selection of the cell or tissue donor and the criteria for excluding the deceased donor, including the objection against the collection of tissues, must be clearly defined. Standard operating procedures should describe the method of cell or tissue collection, detailing the data necessary for the identification of the entity arranging the tissue collection and the data regarding the collection itself so as to ensure full traceability of the collected cells or tissues. The procedures must specify the means of transporting the collected cells or tissues and their acceptance into the tissue and cell bank. The quality system procedures in a tissue bank must specify how to identify and secure cells or tissues during transport, and describe the stage of acceptance of tissue and cellular material into the bank, specifying the information on:

The way in which cells or tissues are accepted into a tissue and cell bank, taking the following into account:

a. The common understanding of the purpose of the collected cells or tissues

b. Results of medical examinations, specialist examinations, and laboratory tests

c. Documented criteria of donor selection performed by an authorized person

d. Documented transport, packaging, and labeling conditions

e. The assessment of the quality of the cells or tissues accepted into the tissue and cell bank

f. The method of dealing with cell or tissue transplants that do not conform with the specification and have incomplete test results and unacceptable quality or defects

g. The course of quarantine from the time of acceptance of cells or tissues into the tissue and cell bank until they are released for processing or withdrawn

The next step, which must be included in standard operating procedures, is the processing of cells or tissues. These procedures must take into account the requirements regarding the following elements of all cell or tissue type processing: labeling, processing conditions, and ongoing evaluation of the processing operations in order to ensure the safety and quality of the cells or tissues being processed. The SOPs must describe the manner of approval and documentation of the changes in the processing operations. A cell or tissue identification system must be developed and described at every stage of the processing to distinguish between authorized products and unauthorized products. The infected cell or tissue removal or withdrawal procedures must be developed and described to prevent infection of other processed cells or tissues, the processing products, environment, or personnel.

The SOPs must include the following:

- The grace period conditions after the processing of cells or tissues
- The method of cell or tissue sterilization
- The grace period conditions after the sterilization of cells or tissues
- The manner of storing the processed cells or tissues and releasing them for distribution after the grace period

Another stage that must be included in the quality system is the storage of cells or tissues after processing. Procedures in this regard must specify the requirements for the following:

- Separate storage conditions and maximum storage time for each cell or tissue type, taking into account the possibility of deterioration of the cell or tissue properties in the course of storage
- Separate storage conditions and maximum storage time for each type of product derived from the processed cells or tissues, taking into account the possibility of deterioration of the product properties
- The development of a cell or tissue inventory and identification system at each storage stage

Standard operating procedures for the distribution of cells or tissues determine the requirements regarding labeling of materials intended for distribution as well as transport criteria and conditions. The scope of standard operating procedures must also include the monitoring of the quality of cells or tissues on their way from the donor to the recipient.

A very important aspect is to define the manner of recording significant adverse events and significant adverse reactions associated with the collection, testing, processing, storage, distribution and transplantation of cells or tissues, their reporting, and elimination of the reasons for their occurrence.

The documentation of the quality system must also contain the operating manuals specifying the applicable documentation, including types of documents, their keeping and circulation in the tissue and cell bank, and the manner of keeping records of specific activities performed in a tissue bank. The documentation must also contain reporting forms and donor cards.

The described quality system concerns procedures regarding tissue and cell banking. If a tissue bank additionally produces advanced therapy medicinal products (ATMPs) using the hospital-exemption advanced therapy medicinal product (HE-ATMP) production procedure, two complementary systems must be developed. Becoming a producer of hospital-exemption advanced therapy medicinal products, the tissue bank must additionally implement the good manufacturing practice (GMP) quality system certified by the Main Pharmaceutical Inspectorate.

The system requires the development of documentation necessary to meet the requirements of the GMP quality system. The basic document containing the producer's characteristics is the site master file describing the activities of the production authorization holder related to the good manufacturing practice.

The GMP documentation must also include the following manuals:

- Specifications describing in detail the requirements for a medicinal product or material used or obtained in the production process. They constitute the basis for quality assessment.

- Production recipes, processing and packaging instructions, and testing manuals. This documentation must provide details of all the output materials, devices, and computer systems used. All instructions regarding the processing, packaging, sampling, and sample testing must also be specified in detail.

In-process control and process control in real time (PAT—process analytical technologies) that are used in the process must be specified along with the acceptance criteria:

- Procedures: standard operating procedures (SOP) determining the manner of operation execution

- Protocols presenting the manuals and records of the concerned operation execution

- Contractual agreements referring to arrangements between the client and the contractor for commissioned activities

The GMP system defines the types of records and reports:

- The records provide evidence of actions taken to demonstrate compliance with the manual, e.g., activities, events, tests, and, in the case of batch production, the history of each batch of a medicinal product, including its distribution. Records contain raw data that are used to create other entries. All data used as the basis for quality activities should be defined as raw data.

- The analytical certificates provide a summary of the medicinal product sample or material test results including the assessment of compliance with a given specification. Alternatively, the certification can be based entirely or in part on real-time process control (summaries and deviation reports) for a batch associated with real-time batch control (PAT—process analytical technologies), with parameters and measurements that must comply with the approved documentation attached to the marketing authorization.

- The reports document the execution of individual tests, projects, or studies as well as results, conclusions, and recommendations.

In the GMP quality system, it is necessary to develop and supervise the documentation, which is defined in the good documentation practice.

It includes the following recommendations:

- The manually entered data must be clear, legible, and impossible to delete.

- The records must be made or supplemented in the course of execution of each activity in such a way that all important activities related to the production of medicinal products are reproducible.

- Every change of a record in a document should be signed and dated, and the means of introducing the change must enable the original information to be read. The reason for the change must be provided in justified cases.

In order to properly archive documents, the relationship between the record and the production activity should be clearly defined as well as the place where the record has been entered.

It is also necessary to have the properly approved and dated specifications of starting and packaging materials as well as the finished products.

The following documents should also be developed:

- Production recipes and production manuals
- Packaging instructions
- Batch production records
- Batch packaging records

The acceptance of each delivery of any starting material (including bulk products, intermediate products, and finished products) of any direct, external, and printed packaging material requires written procedures and records. It is necessary to develop written procedures for material and product testing at various stages of production, describing the methods and equipment used. The executed tests should be documented.

As part of the GMP system, written procedures for releasing or rejecting materials and products, including the certification of the finished products by the qualified person, must be developed. All entries must be available to the qualified person. The developed system must immediately identify deviations and any changes implemented in the critical data.

The distribution records of each manufactured batch of medicinal product must be kept and stored in order to withdraw a series from the market if necessary.

The system also requires the development of written procedures, programs, protocols, reports, and related records regarding undertaken actions or final conclusions. These procedures should include the following:

- Process, equipment, and system validation and qualification
- Device assembly and calibration
- Technology
- Maintenance, cleaning, and disinfection
- Personnel issues, including a list of specimen signatures, and training in respect to the requirements
- Technical issues, protective clothing and hygiene, as well as the verification of the effectiveness of the training sessions conducted
- Environmental monitoring
- Pest control

- Complaints
- Withdrawal from the market
- Returns
- Change control
- Explanatory proceedings regarding deviations and nonconformities
- Internal audits regarding the quality and observing the good manufacturing practice
- Record summaries
- Audits at suppliers

The producers must provide clear and understandable operating instructions for the main production and control equipment. They must keep logs of the master and critical control equipment, production equipment, and areas in which production processes take place. There must be records regarding each area, device, method, calibration, maintenance, cleaning, or repair introduced in chronological order, dated and signed by the personnel performing these activities. A list of quality management system documents should be kept.

4. Quality and risk management in compliance with GMP and with reference to RBV

As described above, the cells were qualified by the European Medicines Agency as advanced therapy medicinal products (ATMPs), so their production is subject to good manufacturing practice (GMP) [17]. The ATMP implementation is therefore not only technologically complex but also strictly regulated by national and European laws. GMP, although it is the legal basis, is only a part of quality management at tissue banks [18]. By definition, resource-based view contains all the principles of good manufacturing practice. If the quality of the delivered product is defined by us as meeting the requirements and expectations of patients, simple observation of the GMP does not guarantee that these requirements will be met [18]. It seems that only the combination of GMP with the ISO 9001 standard, the RBV approach, and lean management enables a comprehensive approach to management, including quality and risk management. ATMP production is special and requires a rigorous and carefully monitored bioprocess to control the intrinsically complex and variable nature of the substance, especially since some ATMPs can be combined with medical devices, such as biodegradable matrices or scaffolds [19]. It should also be mentioned that material collection from a patient takes place in a hospital that is not a GMP-controlled environment, which additionally contributes to the increased variability of this stage [20] and leaves room for other management techniques, such as RBV and lean. A better understanding of graft production management requires risk analysis, including its identification, assessment, and control. The risk is determined by the likelihood of damage and the consequences of this damage. The risk associated with the quality of the advanced therapy medicinal product/tissue is one of the components of the total risk arising from the production and use of the graft. This indicates the need to extend quality management with a resource approach. Product quality must be maintained throughout the entire production cycle and must allow identification and control of potential risks associated with development, production, and transplantation. The quality risk management principles include development, production, distribution, process reviews, and validations. An important element of the quality system resulting from GMP is the process of corrective and preventive actions (CAPA), which refers to deviations, i.e., events departing from the approved procedures or instructions. An example would be a change of the culture incubator during the process or allowing the tissue to be packaged without obtaining the results of quality control analyses. These types of activities are sometimes undertaken to rescue the medicinal product (graft) or to maintain the continuity of cell culture; however, the area of their application ends at the stage of production of the medicinal product. The initial stage in the CAPA process is to clearly define the problem, i.e., register the deviation and provide its short description containing the scope and area of the occurrence of the irregularity and to indicate the leader of the explanatory actions, who will be responsible for carrying out the entire explanatory procedure. At this stage, it must be added that the idea of corrective and preventive action implementation is not to blame anyone but to continuously improve the production process and the quality of grafts. In determining the actual or most probable cause of an adverse event, the following elements should be taken into account: the equipment and materials used, validity, feasibility and comprehensibility of procedures, the design of the entire process (including its bottlenecks), the level employee qualifications, software, and external factors. Any additional resources that may potentially have an impact on the adverse event that has occurred are also identified and documented. After preparing a list of probable causes, the information and data collection must ensue to be used to draw conclusions about the possible cause of the event. Then, the explanatory actions are implemented and the reasons for the deviation are identified. Repair/corrective actions and their implementation are also specified, and all these actions should be described in order to later verify the effectiveness of the steps taken. Then, there is risk communication, which is a process of sharing risk information and risk management methods between decision-makers and other parties. The parties exchange information at each stage of the risk management process. If the explanatory actions do not show the reason for the deviation, a quality risk assessment should be performed. A number of commonly accepted and well-defined methods and processes of risk analysis and quality risk management have been developed. Risk control in turn involves actions that introduce decisions in the area of risk management. The purpose of risk control is to make decisions, which lead to risk reduction to an acceptable level. The contribution of work devoted to risk management should be proportional to the risk weight. Preliminary hazard analysis (PHA) is an analytic tool based on the application of previous experience or knowledge about the threat or failure to identify future threats (Annex 20 GMP). One of the PHA variations is brainstorming, during which the expert group asks "what if" questions to identify the impact of individual elements on the production process and formulate recommendations for the actions to be taken. The quality of the results obtained using this method depends to a large extent on the experience and knowledge of the participants [20]. The hazard and operability studies (HAZOP) method of analysis was developed in the 1960s. Like the PHA, it is a systematic method; however, it requires more detailed information. The HAZOP method uses a predefined set of guiding words to describe the parameters, which leads to the creation of a pair of words that is referenced to a point in the process that can potentially fail. As a result of using this method, a table is created that includes situations, which can cause a failure, together with its consequences and specific causes. However, this is a time-consuming and labor-intensive method and as such it generates considerable costs [20]. The process map is a technique based on a graphical representation of the functioning of a set of processes and their mutual relationships. The fault tree analysis (FTA) has been developed for the aviation industry and is a deductive method that assumes the occurrence of a defect in process functionality and can link multiple causes to identify the cause-and-effect chain. Failure mode and effects analysis (FMEA) is mainly focused on the optimization of the product and is particularly recommended in the situation of new product introduction because it allows for the recognition of the potential interfering factors. As a result of this quantitative method, we obtain the so-called risk priority number (RPN) and information about strong and weak points of production. The criticality of the defect is calculated, and the higher the calculated parameter, the greater the risk associated with the defect. After identifying the risks and weaknesses of the process and their characterization, decisions should be made on which risks should be reduced and which should be observed or eliminated. Controlling risks is a technological and economic challenge. Before making any changes, an assessment must be performed to ensure that the proposed change will not cause any new or unexpected risks. The fact that a change is inevitable makes it a critical factor, especially in GMP and ISO environments, where inappropriate or "uncontrolled" changes can affect patient safety and public health. For this reason, the concept of change control is closely related to compliance with GMP and ISO, where any changes in production and processes must be controlled. Change control procedures must be recorded in order to standardize the workflow, especially at key stages such as collecting material outside the production environment. The "uncontrolled" change refers to modifications made without verification and approval by the quality control manager and in special cases also the hospital management. In GMP and ISO environments, strict adherence to approved policies and procedures is a key factor in maintaining production efficiency in a controlled state, and change control is critical. Changes are subject to review and approval by the quality control unit. To summarize, GMP is a formalized procedure and imposes a heavy burden on the producer of transplants, and while compliance is a required minimum, it is not enough to ensure the right quality of using the manufactured products [18]. The specificity of risk management in the aspect of GMP consists in focusing on potential failures in the production of transplants and their safety for the patient. RBV refers to a broader management area including risk management at the time of collecting cellular material and its transplantation as well as the patient's fate. However, some authors point to limitations in the application of resource management in the public health service [8] due to the complex nature of such entities. Lean management possesses tools that are able to systematize it [21].

5. Production management

The tissue bank is the place for graft preparation for the treatment of patients with severe burns and chronic wounds. As already mentioned, the preparations produced in our bank are biostatic grafts, which are subject to radiation sterilization and live cell transplants, produced under sterile conditions. Due to the specificity of the products manufactured in the laboratories, appropriate, supervised environmental conditions must be ensured. For this purpose, we use "clean rooms," in which, thanks to the use of special HEPA filters and laminar air vents, it is possible to obtain the appropriate class of air purity. The highest class achieved in our bank is air purity class A achieved under the laminar chamber, where the amount of particles and microorganisms generated during operation should be zero. In order to provide such sterile and dust-free conditions in air purity class A, the rooms located on the way to a class A clean room must meet a series of criteria. The main one is the maintenance of a pressure cascade, which means that in order to get to a class A clean room, you have to go through several airlocks separating neighboring rooms of the following purity class: gray (of the least air purity) D, C, and B. Positive pressure is maintained in each airlock in relation to the previous lock, employees wear dust-free and sterile clothes and gloves, and, finally, in the class B air purity room, they put on a sterile dust-free clean room suit covering their entire body. All these precautions are necessary because we want to protect the product we manufacture, i.e., the advanced therapy medicinal product in the form of cultured skin cells. The cultures that we prepare are supervised by the Main Pharmaceutical Inspector; therefore, as mentioned above, they must be prepared in accordance with good manufacturing practice (GMP). The requirements that we must meet in order to culture cells for a burn patient make all the preparations leading to the process initiation strategic. It should be noted that employees and resources (reagents and consumables) in the laboratory are in a constant state of readiness. At the same time, we are not able to estimate with 100% certainty whether, at a given time, we are going to culture cells at all and, if so, for how many patients. The unpredictability of the production process (cell culture) in the face of the risk of expiry of reagents necessary for maintaining production continuity means that a compromise must be found consisting in the continuous maintenance of small-scale stock of culture materials. The greater part of consumables used in the preparation of medicinal products is highly specialized and is not widely available, and the delivery of a larger quantity often takes a long time. Therefore, ensuring constant availability of materials seems to be one of the solutions enabling continuity of production. However, lean management requires keeping losses to a minimum, which is, in a way, contrary to the GMP assumptions making production halt risk minimization a priority. A common-sense approach should therefore be applied translating into ensuring the minimum amount of reagents that is sufficient to maintain the continuity of production, even in the case of mass events, such as an explosion in a coal mine, when mass production of cellular grafts must be available immediately.

In the case of maintaining sterile environmental conditions, it is also important to systematically and periodically perform cleaning and disinfection of rooms according to the adopted schedule. It is also necessary to permanently perform environmental monitoring: continuous monitoring of the amount of particles in class B rooms, as well as microbiological tests of air and clean room surfaces, carried out according to schedule. Control and supervision are also applied to devices used for cell culture and the conditions prevailing in them, e.g., the devices that are critical in the production process—incubators. A monitoring system is installed in the rooms. It monitors the environment and notifies employees supervising cell cultures about the occurrence of errors in the culture process. Such monitoring is necessary because it allows for quick response and taking action in the event of a risk of culture loss.

Highly specialized persons who have acquired appropriate skills in numerous training sessions are assigned to work in the clean rooms. Their professional experience and the ability to organize and manage their working time are also important. Due to the specifics of the work—a sterile, monitored and supervised environment in a clean room—each entrance to such rooms generates additional costs. That is why, employees working with cell cultures must carefully consider, plan, and organize all elements of their work in the laboratory before it begins. Employees involved in the preparation of cell cultures follow the applicable, written procedures and validated processes. Most processes should be validated based on reference reagents; however, the tissue material from which the cells are isolated and the culture is established is so unique that it is not possible to replace it with the reference material. Therefore, all employees must comply with specific requirements. In standard work, it is important not only to maintain the purity and sterility of the cell culture and the rooms but also to ensure that all necessary reagents and materials are available in the laboratory. The duty of employees leaving the laboratory is to supplement, prepare, or provide information on missing items to all persons involved in the cell culture process. Each employee entering the laboratory studio must be sure that he or she will be able to perform all the tasks without any problems. Due to the need to monitor and maintain room sterility, any unnecessary entering and exiting the clean room creates a risk of pollution and generates costs of, among other things, used protective clothing and cleaning agents. All these activities are aimed at ensuring the sterility of the advanced therapy medicinal product manufactured in the tissue bank. **Figure 2** shows the increase in the number of cultures and autologous skin cell transplantations. This increase has been possible thanks to the standardization of production processes that has been achieved by minimizing material losses. The lean and TOC management methods contributed to the increase in the number of cultures, which made better management of laboratory resources possible.

As mentioned above, cell cultures can be established after the occurrence of a mass event, e.g., an explosion in a coal mine or a large fire, in which more people suffer burns. In such a situation, our only safeguard that makes it possible to take appropriate action is to keep reagents and consumables in the laboratory. In such cases, the experience of employees and cooperation with other hospital departments are very important. This allows for the planning of a strategic approach to the problem, which goes beyond the GMP management area, and still requires management and control. Establishing a cell culture is conditioned not only by securing the necessary reagents and materials but also by providing care for the most disadvantaged patients, which results from the efficient operation of the hospital, reinforced by the management system. In such cases, it is also important to adopt the right approach and plan the work of all personnel to avoid unwanted cross-infection or contamination, which could result in the loss of valuable cell culture. The resource-based view (RBV) approach seems to be necessary to manage the aspects that are not covered by the GMP procedures.

The preparation of biostatic skin and human amnion grafts is much less restrictive. These are grafts, in which the final stage of preparation is radiation

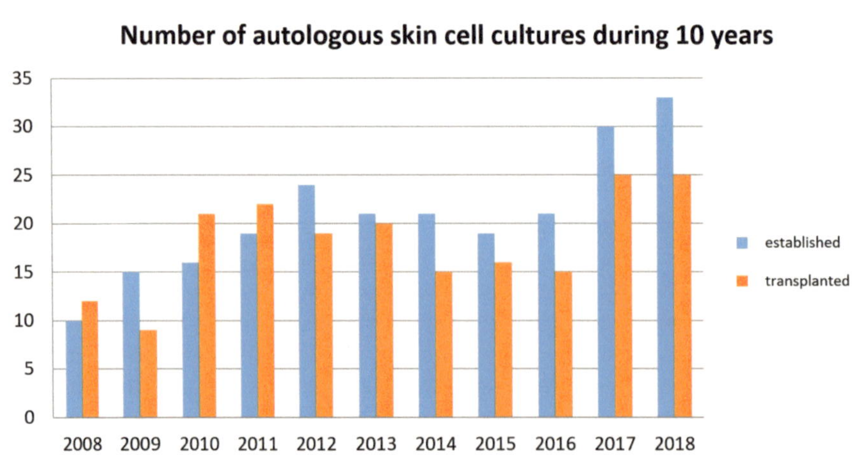

Figure 2.
Number of cultured and transplanted autologous skin cell cultures in years 2008–2018.

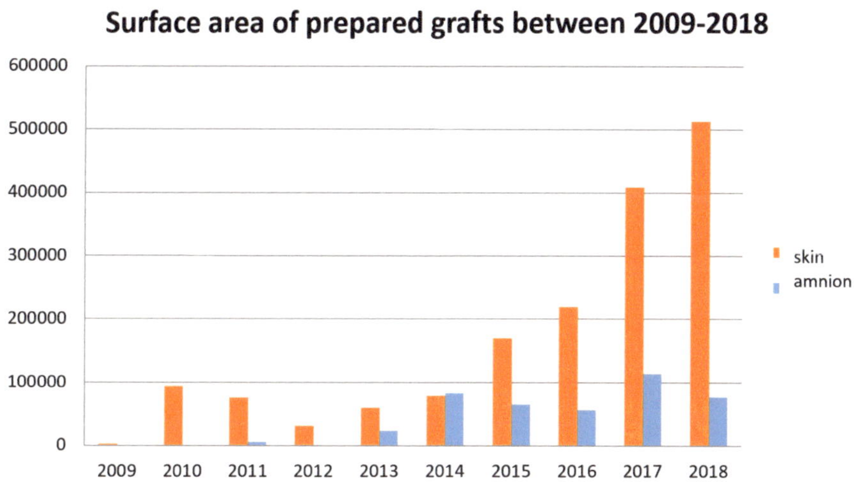

Figure 3.
The surface area of prepared skin and amniotic grafts in years 2009–2018.

sterilization. This means that these transplants are not live and constitute only a temporary dressing. Due to the fact of sterilization, the grafts are also prepared in clean rooms, under the laminar chamber of class A air purity, but in the class C environment. When preparing tissue transplants, employees also use their experience in managing their working time, maintaining cleanliness and order in the laboratory and good manufacturing practice. These skills are very useful and promote continuous improvement of the produced grafts and production efficiency. The chart in **Figure 3** shows the increase in the surface area of biostatic amniotic and dermal grafts prepared in particular years. The use of a resource management approach, among others, has enabled the increase in productivity, which translates into the surface area (cm^2) of the prepared tissue and cellular grafts. The increasing number of donors, and hence the amount of documentation to complete, makes it necessary to adopt a strategic approach to the work so that as much work as possible can be performed at the same time, thus generating added value.

6. Strategic potential of a tissue bank in terms of validation of processes and qualification of equipment and rooms

Tissue bank management requires well-established, thorough knowledge of the processes occurring in the area of clean room production. In this respect, the key issue for a tissue bank, as mentioned above, is to have adequately qualified personnel and therefore human resources possessing the required education, experience, competences, and abilities. The second important element in the process of managing a tissue bank is the possession of appropriate devices and rooms, i.e., the hardware and accommodation resources. User requirements for hardware and accommodation resources should be characterized and defined in detail. These elements, in the RBV approach, can be collectively referred to as strategic potential. In general terms, the strategic potential is a factor influencing the tissue bank's achievement of the expected results. The key effects include the ability to use resources to achieve the intended targets and, therefore, obtain the highest quality product according to the tested, repeatable, and effective production methodology using adequately verified equipment and rooms.

The expected effects are, therefore, the result of the tissue bank's implementation of adequately planned tasks. These tasks are directly related to the achievement of the intended objective. In this respect, the achieved production effects are the so-called determinants of a tissue bank success. We use this expression to appropriately define the ranges and criteria of acceptance that a given tissue bank product must meet before a batch is released to distribution. These criteria must be strictly defined in the production specification. Moreover, in addition to the so-called determinants of success, so-called success factors must also be defined. These factors result from adequately defined and purposefully adopted strategic potentials of the tissue bank. The strategic potentials include in particular the professional experience of human resources. Based on specific experience and "know-how," appropriate operational procedures should be created for all activities performed in the tissue bank. However, the final adoption of all the proposed solutions (e.g., methodological, analytical, and technological) requires evaluation that is very carefully planned and analyzed in terms of the correctness of the adopted assumptions. This means that each parameter that constitutes a specific stage or the whole of a given process must be subject to control in terms of meeting the expected assumptions. In case of changes in the specified parameter or parameters, a whole evaluation of the process must be repeated.

Procedures related to the evaluation of processes executed in the tissue bank are called validation procedures. In this respect, special consideration should be given to the validation of the manufacturing process of the hospital-exemption advanced therapy medicinal product (HE-ATMP), the process of manufacturing medical devices, tissue transplants, transport of these products, cleaning, and disinfection, including employee clothing. However, by contrast, the evaluation procedures related to the correct functioning of devices and rooms are called qualification procedures. In this respect, it should be clearly emphasized that the qualification of devices and rooms covers the entirety of undertakings related to the purchase, installation, and operation of a given device or room, starting from specifying the user's specific requirements to determining the conformity of the proposed preliminary design with the requirements, to subsequent processes confirming the compliance regarding the installation and proper functioning.

6.1 Main validation plan

The main validation plan (MVP) defines the approach to the qualification/validation of GMP areas for all validation activities carried out in the existing or newly designed premises of a tissue bank. It is aimed at ensuring that the expectations have been clearly communicated to all participants implementing the validation program. The MVP clearly and comprehensively defines the requirements and scope of responsibility in the validation process. The main objectives and tasks of the MVP are as follows:

1. Presentation of policy, requirements, and expectations regarding the validation activities in laboratory rooms where the tasks related to the production of advanced therapy medicinal products, HE-ATMP, analysis and quality control, as well as research and development work are performed

2. Defining the organizational structure of validation activities

3. A brief description of the installations, systems, and devices that will be validated with reference to the existing documents

4. Specification of the format of the validation documentation used in the protocols and reports

5. Outlining the validation strategy and indicating the documents defining the task and scope of responsibility for the tissue bank, contractors and suppliers of particular systems, installations, and process equipment

6. Presentation of the approach to change control

7. Drawing up a plan and scheduling validation activities

8. Ensuring consistent application of the terms in accordance with their established definitions (in documents such as the standard operating procedures (SOPs), validation protocols, etc.)

The MPV contains general guidelines for the preparation of other validation documents. Rational principles and a rational approach when using the MPV are recommended. In the event of circumstances unforeseen in the plan, the rationale and assessment of the deviation must be documented.

Validation/qualification activities will be carried out depending on the subject of validation/qualification and on the nature of the activities performed by the validation team appointed by the qualified/competent/responsible person to perform specific tasks. The validation team may consist of representatives of quality assurance, tissue bank user/manager, technical and technological experts, and quality control.

This approach to validation and carrying out validation work fulfills the regulations and guidelines specified in the European Commission Directive 2003/94/EC and the Regulation of the Minister of Health of October 1, 2008, on the requirements of good manufacturing practice (Journal of Laws 2008.184.1143) as amended. Validation is an action aimed at confirming, in a documented manner and in accordance with the principles of RBV and good manufacturing practice, those procedures, processes, devices, materials, activities, systems, and installations truly lead to planned results. This is achieved through the development of test plans, protocols, and procedures, as well as the implementation of the records made in the protocols and the documentation of the results obtained in the intermediate and final reports. The report is an approved written plan of the measurement, control, and methodology of testing and result documentation. The verification and validation of the qualification and validation documentation are performed in accordance with the internal procedure prior to the initiation of the qualification/validation activities. In any case, the validation activities performed by external companies must be supervised by an appropriate employee of the tissue bank and are subject to acceptance and approval by tissue bank employees. The same persons and the persons performing the tests (other than those included in the validation team) also check and approve the reports from a given qualification/validation phase following the tests, in order to confirm that the production environment, devices, and the process are indeed suitable for the production of HE-ATMP, medical devices, tissue grafts, etc., in the tissue bank and that they meet the requirements of good manufacturing practice. The full set of qualification and validation documents consists of protocols and test cards.

The installation qualification (IQ) protocols pertaining to the equipment, installations, and clean rooms, similar to the operational qualification (OQ) protocols of these systems, are developed for all critical modules of systems, installations, and devices.

The process qualification (PQ) protocols are developed only for those systems, installations, or devices for which their operation data are necessary to verify the correctness of the process and for which long-term monitoring of their performance is recommended (e.g., laminar chamber, CO_2 incubator, bioreactor, and refrigerator-freezer). Systems, installations, or devices for which PQ activities are necessary will be selected at the risk analysis (RA) stage. Any noncompliance with the conditions set out in the protocols encountered during the validation process will be recorded and evaluated. The registered nonconformities will be analyzed, and then the actions regarding the manner of further dealing with the nonconformities (e.g., the development of justification/explanation, making corrections, and then carrying out reclassification tests) will be defined. All such data will be documented and saved. The obtained results will be collected and presented in relevant reports after the completion of each qualification stage.

6.2 Approach to validation

Validation of rooms, devices, systems, and installations used in production carried out in the tissue bank constitutes a part of the quality assurance policy of the entire enterprise (tissue bank, hospital, and biotechnological or pharmaceutical company). Qualification/validation activities and the creation of appropriate documentation must be carried out in accordance with the qualification/validation schedule by tissue bank representatives, suppliers of installations, systems, process equipment, and/or contractors providing validation services. Supervision and coordination of qualification/validation activities should be carried out by a tissue bank representative. Validation teams and validation contractors are required to develop a series of protocols to check all critical parameters of rooms, systems, installations, and devices in the tissue bank. The protocols will be implemented in accordance with the test methods presented in the protocols. Each protocol will be used to obtain documentation that will confirm the compliance of the system, installation, or device with the GMP guidelines. To qualify equipment, rooms, and installations, the documents obtained after carrying out the service activities can be used. All protocols and reports will be archived after their final approval.

It should be emphasized that conducting validation cannot negatively affect the production process. In addition, special attention should be paid to the timely performance of validation work, and subsequent planned validation activities related to a given device, room, etc., should be carried out within the prescribed time limit. However, it is possible that the deadline of the next validation is not met. This disadvantageous situation may occur due to a number of possible occurrences that may affect the feasibility of performing validation work. The factors affecting the performed validation work time shift include, among others, production, sickness of the person performing the validation, mechanical failure, etc. The information about the validation time shift must be included in the validation protocol.

In order to summarize and approve the completion of individual stages of qualification/validation activities, appropriate reports are prepared, which are checked and approved by appropriate persons. The report must include a summary of the qualification/validation course, a description of test results, and documented, explained, and evaluated nonconformities. After the approval of the report, if no critical nonconformities were found or all critical nonconformities were resolved, you are allowed to proceed to the next qualification/validation phase.

6.3 Summary

The activities related to the performance of validation/qualification activities are aimed at ensuring the highest quality of the tissue bank products being developed. They take place in accordance with RBV and with the optimal use of the strategic potential of the laboratory. The approach to validation is part of the broadly understood RBV, which results from the conscious management of the unit using knowledge and science. The validation and qualification activities are necessary to verify the adopted methods and the continuous improvement of processes. This, in turn, fits into modern management systems using the PDCA or DMAIC approach, based on a thorough analysis of a given process and a thorough problem solution. It should also be emphasized that the validation activities require the involvement of appropriately qualified personnel, which will contribute to ensuring high quality and safety of the obtained products.

Author details

Wojciech Smętek[1*], Jacek Węgrzyk[1], Agnieszka Klama-Baryła[1,2], Wojciech Łabuś[1], Małgorzata Kraut[1], Michał Szapski[1], Mariusz Nowak[1] and Diana Kitala[1,2]

1 Stanislaw Sakiel's Center for Burn Treatment, Siemianowice Śląskie, Poland

2 Silesian Higher Medical School in Katowice, Katowice, Poland

*Address all correspondence to: fundusze@clo.com.pl

References

[1] Rummler G, Brache AP. Improving Performance: How to Manage the White Space on the Organization Chart. 3rd ed. Hoboken, New Jersey: Wiley; 2012

[2] Jimmerson C. Value Stream Mapping for Healthcare Made Easy. Boca Raton, Florida: CRC Press, Taylor & Francis Group; 2009

[3] Goldratt EM, Cox J. The Goal: A Process of Ongoing Improvement. 3rd revised ed. Great Barrington, MA: The North River Press Publishing Corporation; 2004

[4] Cox JF, Schleier JG. Theory of Constraints Handbook. New York: McGraw-Hill; 2010

[5] Protzman C, Kerpchar J, Mayzell G. Leveraging Lean in Ancillary Hospital Services. Boca Raton, FL: CRC Press; Taylor&Francis Group; 2015

[6] Das TK, Teng B-S. A resource-based theory of strategic alliances. Journal of Management. 2000;**26**(1):31-61

[7] Fahy J. The resource-based view of the firm: Some stumbling-blocks on the road to understanding sustainable competitive advantage. Journal of European Industrial Training. 2000;**24**(2/3/4):94-104

[8] Ferlie E. Resource based view: A promising new theory for healthcare organizations: Comment on "Resource based view of the firm as a theoretical lens on the organisational consequences of quality improvement". International Journal of Health Policy and Management. 2014;**3**(6):347-348

[9] Barney JB, Clark DN. Resource-Based Theory: Creating and Sustaining Competitive Advantage. Oxford: Oxford University Press; 2007

[10] Barney JB. Organizational culture: Can it be a source of sustained competitive advantage? The Academy of Management Review. 1986;**11**(3):656-665

[11] Hamel G, Prahalad CK. Competing in the new economy: Managing out of bounds. Strategic Management Journal. 1996;**17**(3):237-242

[12] Dickson PR. The static and dynamic mechanics of competition: A comment on Hunt and Morgans comparative advantage theory. Journal of Marketing. 1996;**60**(4):102-106

[13] Barney J. Firm resources and sustained competitive advantage. Journal of Management. 1991;**17**(1):99-120

[14] Chatterjee S, Wernerfelt B. The link between resources and type of diversification: Theory and evidence. Strategic Management Journal. 1991;**12**(1):33-48

[15] Yarbrough AK, Powers TL. A resource-based view of partnership strategies in health care organizations. Journal of Hospital Marketing & Public Relations. 2006;**17**(1):45-65

[16] Madhani PM. Resource Based View (RBV) of Competitive Advantage: An Overview. Hyderabad, India: Icfai University Press; 2010

[17] Kitala D, Klama-Baryła A, Kraut M, Łabuś W, Kawecki M. Isolation, culturing and preparation for transplantation of amniotic mesenchymal stem cells: Repetitive and reproducible laboratory, technical protocol. Annales de Biologie Clinique (Paris). 2018;**76**(5):562-567

[18] Willig SH. Good Manufacturing Practices for Pharmaceuticals: A Plan for Total Quality From Manufacturer to Consumer. New York: Marcel Dekker; 2001

[19] Dream R, Odum J. Impact of ATMP manufacturing on process equipment and facility design. BioPharm International. 2018;**31**(11):30-34

[20] Vesper JL. Assessing and managing risks in a GMP environment. BioPharm International. 2005;**18**(3):46-58

[21] Lawal AK, Rotter T, Kinsman L, Sari N, Harrison L, Jeffery C, et al. Lean management in health care: Definition, concepts, methodology and effects reported (systematic review protocol). Systematic Reviews. 2014;**3**(1)